The geography of contemporary China

Deng Xiaoping's rule has brought fundamental economic and social changes to China. In this book a group of specialists on China consider the impact of these years on China's people, industry, agriculture, trade and environment. The contributors assess a decade of Chinese policies towards regional development, urbanisation and geopolitics. The book also contains a postscript which outlines the emerging trends and policies since the events of June 1989 and indicates the resulting shifts in geographical priorities.

Readable and accessible, this introduction to the geography of Communist China sets the country's spatial and environmental problems in their historical, political and economic framework. Designed for students of geography and allied subjects, the book is also suited to specialist courses on China and to development courses which use case studies of China's development strategies.

The geography of contemporary China

The impact of Deng Xiaoping's decade

Post-mao Decade

Edited by

Terry Cannon
and
Alan Jenkins

London and New York

First published 1990
by Routledge
11 New Fetter Lane, London EC4P 4EE

Simultaneously published in the USA and Canada
by Routledge
a division of Routledge, Chapman and Hall, Inc.
29 West 35th Street, New York, NY 10001

Typeset by J&L Composition Ltd, Filey, North Yorkshire
Printed and bound in Great Britain by Mackays of Chatham PLC, Kent

British Library Cataloguing in Publication Data

The geography of contemporary China : the impact of Deng
  Xiaoping's decade.
  1. China. Political events
  I. Cannon, Terry II. Jenkins, Alan
  951.058
IBSN 0–4415–00102–1

Library of Congress Cataloging in Publication Data

The Geography of contemporary China : the impact of Deng Xiaoping's
  decade / edited by Terry Cannon and Alan Jenkins.
      p.   cm.
  Includes bibliographical references.
  ISBN 0–415–00102–1. — ISBN 0415–00103–X (pbk.)
  1. China—Economic policy—1976–   2. China—Economic
conditions—1976–   3. China—social conditions—1976–   I. Cannon,
Terry, 1948–   . II. Jenkins, Alan.
HC427.92.G46   1990
338.951′009′047—dc20
                                                            90–8271
                                                            CIP

Contents

Figures and tables

Tables

Contributors

Terry Cannon is currently on secondment as lecturer to the Institute of Social Studies, The Hague. He was previously at the School of Humanities, Thames Polytechnic in London, where he teaches Development Studies. He has published articles and chapters on China's regional development, national minorities and foreign economic relations, and until 1989 was editor of the magazine *China Now*.

Edward Derbyshire is Professor of Geography at the University of Leicester. He is Director of the Centre for Loess Research and Documentation, and Senior European Community Consultant on research and control of landslides and debris flows in Gansu Province, China. His main reseach interest is loess and glaciation, especially in high Asia.

Richard Louis Edmonds is a lecturer in Geography with reference to China at the School of Oriental and African Studies, University of London. His interests are historical geography and environmental problems in China and Japan. He is author of *Northern Frontiers of Qing China and Tokugawa Japan: a Comparative Study of Frontier Policy* (1985, Chicago: Chicago University Press).

Bernhard Glaeser is a project director at the Wissenschaftszentrum Berlin für Sozialforschung (WZB). Research interests include human ecology and environmental policy in developing countries. He led a research project on Chinese environmental policy and edited *Learning from China* (1987, London: Allen & Unwin).

Maurice Howard is an economic advisor at the Department of Transport. He has worked as a consultant on Chinese economic affairs, and is author of a chapter in D. Goodman (ed.) *China's Regional Development* (1989, London: Routledge). He has a special interest in transport, industry and the urban economy.

Alan Jenkins is a principal lecturer in the Geography Section of Oxford Polytechnic. Research interests include Chinese geopolitics and Western views of China, particularly as represented in documentary films. With David Pepper he edited *The Geography of War and Peace* (1985, Oxford: Blackwell).

John Jowett is in the Department of Geography, Glasgow University. He was a research visitor to China in 1975, 1983, and for four months in 1988. His recent research has focused on population problems and policies. Publications on China have appeared in *The Geographical Journal, GeoJournal, Geography* and *The China Quarterly*.

Frank Leeming is a senior lecturer in the School of Geography at the University of Leeds. He is the author of *Rural China Today* (1985, London: Longman) and has also written on Chinese industrial and historical geography.

David R. Phillips is a lecturer in Geography at the University of Exeter. His interests include economic and social issues in Hong Kong and southeast Asia, Chinese housing, trade, and Special Economic Zones. He is author of *China's Modernisation: Prospects and Problems for the West* (1985, London: Institute of Conflict Studies) and co-editor of *New Towns in East and Southeast Asia* (1987, Oxford: Oxford University Press).

Simon Powell is a legislative analyst for the General Assembly of Maryland in the USA. His doctoral research was concerned with rural China. With Behrman and Fischer he has recently written a book *Doing Business with China*.

Wing-Shing Tang is a lecturer in Geography at the Chinese University of Hong Kong, and has researched China's urbanisation and the development of Shanghai.

Anthony Gar-On Yeh is a lecturer in the Centre of Urban Studies and Urban Planning at the University of Hong Kong. He is author of articles on urban development and Special Economic Zones in China. He is co-editor of *New Towns in East and Southeast Asia* (1987, Oxford: Oxford University Press).

Preface

Western knowledge of China is the poorer because it has lacked much geographic analysis, and this book seeks to show the value of such a perspective. Although in some ways it resembles a regional geography text book, it is different. The normal descriptive emphasis, on the location of activities, is there. But it is soon apparent that most of what the various authors discuss is largely explained through analysing changes in the wider political and economic system.

China is a country in which, more than in most others in the world, state policies shape economic and social issues, and hence geography. This is why it makes sense to have the sub-title 'The impact of Deng Xiaoping's decade'. For ten years from 1979, his influence on the Communist Party and government have been highly significant; and he has represented the views of the dominant sections of the party.

It has been a period of enormous changes, contrasting sharply with the policies of the preceding years, dominated by Mao Zedong, whose death in 1976 signalled the start of the changes. The book is mainly aimed at understanding the last decade, while providing newcomers to China with sufficient information about the period 1949 to 1976 for them to be able to recognise the significance of the changes.

Because of the overwhelming significance of the policies of the government in determining the geography of China, writing a book on the country is a hazardous business. By the time it is finished it is usually time to start it all over again, because the policies alter or other events change matters quickly. For example, during the writing of this book, in March 1988 the National People's Congress (the 'parliament') amended the 1982 Constitution to allow the sale and resale of 'land-use rights'. This drastically alters the balance of control of land, previously declared to be a public good, in favour of private control.

This is the sort of policy shift which people who study China have become very accustomed to in the last ten years. As we finished preparing this book for publication in 1989, the extraordinary May mass demonstrations were developing. We already felt that something was happening which potentially made the book not so much out of date as left behind by events.

Far worse, and much more difficult to accommodate, were the massacres of June 1989. These showed the inability of the old party and the army leadership to cope with the political dissent which grew from the negative effects of the economic reforms of the last decade, and the lack of progress towards democracy. The implications of those events for the reform policies cannot yet be understood. Although we have been able to rewrite the Introduction in July 1989, we must stress that most of the book was written before that, and the various authors have not had the opportunity to update their contributions. However, we, as editors, have had the chance to add a Postscript (p. 291) which assesses the geography of policy changes since June 1989. We urge all to read it as a first stage of understanding the new situation.

But books have eventually to go to publishers, so by the time this is read, whatever the validity of what we wrote, it may have been left behind; policies may have changed again. The book remains as a survey of forty years of Communist Party rule, with the emphasis on the last decade of Deng Xioaping's leadership. But beware: use the sources at the end of the Introduction (Chapter 1) and some other chapters to 'update' what has been written. These Update sections enable you to find material which will help you understand what has happened in policy or other events in relation to particular themes.

There are other reasons why readers should be aware of the partiality of this book. The authors were chosen for their particular specialist knowledge of China as geographers, not for some preferred political analysis. Readers will note differences in their interpretation. That is to the good, for it emphasises that the perception of reality in China (or elsewhere) is shaped by one's particular viewpoint. But the authors are from Britain, Germany and Hong Kong. Does that shape their analysis? Would a text by particular Chinese geographers have greater validity or be less partial?

Furthermore, you will note that many chapters are dependent on Chinese sources. Yet several chapters indicate that such sources, for instance on the Great Leap Forward (1958–60), were at best propaganda in which the Chinese misled themselves, and at worst deliberate misrepresentation. Most authors have used recent government yearbooks and official publications. We think these are very different from the information available before 1979. In our view the media has been far more informative and in our terms more 'honest' during the past decade. But we would argue that official sources should be used with care, no matter what their provenance; Western governments also are very capable of producing misleading information and manipulated statistics. The problem with China is the more obvious after the 1989 massacres, given the blatant attempts of the government to pretend to the people that nothing happened. There are other gaps in the book. We chose to exclude any treatment of Hong Kong and Taiwan, except as related to the content of relevant sections like Chapters 9 and 11. Our reason for this is mainly that our focus is not a traditional region, but the entity of the People's Republic of China. It is a book largely about the connections of the geography of that state with the political, social and economic changes which emerged from the establishment of the communist-led government there after 1949, and the impact in turn of the territory's geography on the policies of that period.

The book has its origins in the activities of a Working Party of the Institute of British Geographers which, with its informal China Geography Network counterpart, brought together many British geographers with an interest in the country. One of the main aims of these groups was to promote the teaching of China in geography, and increase the number of courses related to the largest nation on earth from the very few existing in Britain. To help do this, it was recognised that more teaching materials were needed, and this book is one of the results.

There are very few geography text books available on China, and to our knowledge there are no others in English aimed at degree level published after 1983. We hope that this contribution will make it possible for many teaching and learning geography in higher education and upper schools to do some work on China.

The book may be used without regard to the chapter order, although there is a logic to the progession of most chapters. Our only recommendation is that should you feel that your knowledge of the country is limited, then Chapter 1 (Introduction) ought to be read first. It provides a guide to the main policy changes of the last decade by contrast to the period from the setting-up of the People's Republic in 1949 to the death of Mao Zedong in 1976. This chapter actually says very little about geographical issues. Spatial analysis is first dealt with, at the macro level, in Chapter 2.

Almost all chapters contain a case study related to one of its main points of analysis. We have encouraged these to be provided because the general dearth of knowledge about China means that it is easy for many people to be rather overwhelmed by the enormity of the task of getting to understand it. The case studies provide a means for readers to become more familiar with at least one smaller-scale aspect of the country, and to feel less intimidated by the large-scale treatment which is inevitable for the substance of the book.

This problem of getting to grips with the 'giant of Asia' is also helped, we hope, by the provision of some basic data about the country, a glossary, and a chronology. There are also short lists of further reading at the end of each chapter. These include some suggested by the authors, but have been augmented by us, the editors, especially to include books which have been published since the chapters were written. These are intended to be of direct use in follow-up work by students needing more detailed information on a particular theme. To improve the readability of the book, most chapters do not include references in the text, especially when these are not widely available or are in languages other than English.

We hope that the book will do a job in stimulating much wider interest in China, a country which deserves to be much more generally known, and which despite the repulsion created by the 1989 massacres is going to have an increasing role in the world. Its people should not now be ignored or abandoned as they have for too long by geography and geographers in some English-speaking countries.

Our appreciation goes to the many people who have helped make this book, especially the contributors. They have also shown patience in waiting for it all to come together. We thank the IBG for its support of the short-life Working Party forming the basis of meetings and seminars which generated papers and discussion from which the book emerged. We are also grateful to our partners and children for their support and forebearance of the disruption to domestic life which the production of the book brought on them.

Terry Cannon and Alan Jenkins,
July 1989

If you are uncertain about the geography of China, take heart!

In 1929, revolutionary armies were sweeping through south and central China. For the first time, perhaps, since 1900 the Chinese crisis became the immediate concern of the [British] cabinet. Mr Baldwin [the Prime Minister] visited a room in the Foreign Office, in company with the Foreign Secretary, Mr Austen Chamberlain. They were shown the large scale map of China. 'So Canton is down there', exclaimed the Prime Minister, 'I always thought it was here' – pointing to the neighbourhood of Tientsin. Mr Chamberlain contributed the remark that China was really very much larger than Japan.

Having identified the position of China on the map, and appreciated the importance and political influence of the British interests here, they decided to send a strong force of troops to garrison the international settlement.

Source: Fitzgerald, C. P. (1964) *The Birth of Communist China*, Harmondsworth: Penguin, pp. 64–5

Basic information

Land area: 9,600,000 square kilometres

Population: 1,100,000 reached in 1989

Population growth rate: average of 1.2 per cent per annum 1981–85

Gross National Product (GNP): RMB 788,000 million in 1985

GNP per capita: RMB 754 in 1985, or approximately US$ 400, putting China towards the top of the poorest third of nations

Land use by topography: mountainous 33 per cent; plateaux 26 per cent; basins 19 per cent; plains 12 per cent; hills 10 per cent – all approximate

Land use by function: cultivated 10 per cent; forests 12.5 per cent; fresh water 1.75 per cent; usable prairie 23.4 per cent

Currency: Renminbi (RMB), translated as The People's Currency, the unit of which is the yuan. In 1988, the official exchange rate against the pound sterling was around 1 = yuan 7

CHINESE NAMES

The capital of China is spelt 'Beijing' in some sources and 'Peking' in others. Which is correct? The problem is that Chinese is a language the written form of which is displayed in ideograms or characters. To know how to pronounce them, they have to be learnt by heart. And to avoid having to use them in foreign publications, some means of transcribing them into familiar Roman letters has to be found (see also Case Study 3.1).

The Chinese have devised a system called *pinyin*, which is now widely used in China and abroad, so we have chosen to adopt it throughout this book. However, in Britain another major system has traditionally been

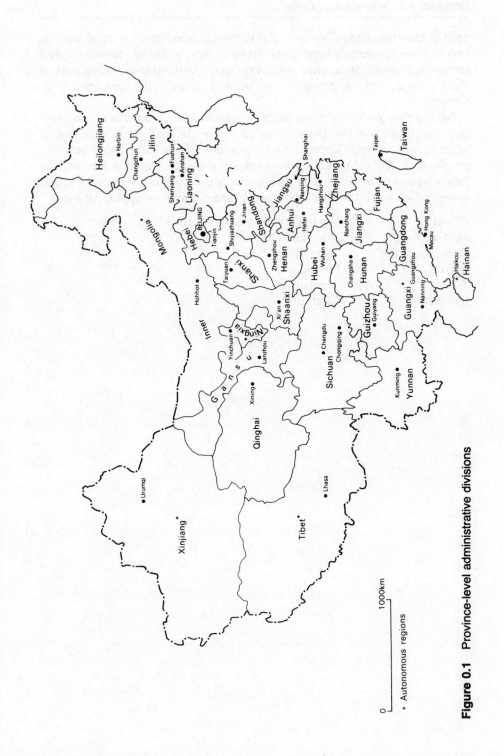

Figure 0.1 Province-level administrative divisions

used in transliterating Chinese – the Wade-Giles method – named after its two nineteenth-century exponents. Some of their words are familiar; other common Roman forms have come into usage without even being part of their system, out of common use of traders over the centuries (e.g. Canton).

Because the *pinyin* versions are now standard, we use them in this book but sometimes put the Wade-Giles or other old spelling in parentheses when it may be well known. As an example of the differences between *pinyin* and the other common ways of showing Chinese names, the following is a short list of some more widely used names. Some people and places which were familiar in the old systems may now appear strange.

Pinyin	**Wade-Giles or other old name**
Mao Zedong	Mao Tse-Tung
Deng Xiaoping	Teng Hsiao-Ping
Beijing	Peking
Tianjin	Tientsin
Guanzhou	Canton

Chronology: from Opium War to Deng Xiaoping

1839 After Customs Commissioner Lin in Canton destroys British-smuggled opium, trade in which the Chinese government is trying to prevent, Britain launches attacks on coastal cities, so beginning the Opium War.

1842 Treaty of Nanking signals China's defeat, and Hong Kong island is ceded to Britain; trading rights granted to foreign nations.

1851–65 The Taiping Rebellion encompasses much of east China, with millions of people killed in the battles and sieges against Imperial targets. Put down by joint forces of Emperor and Western powers.

1860 British and French forces occupy Beijing in war over trade rights. Russia aquires huge areas of Manchuria.

1895 Japan defeats China in war for territory, and takes Taiwan and Liaodong peninsula.

1905 After war between Japan and Russia in Manchuria, Japan takes control of northeast China.

1911 Republic of China established after success of Nationalist Party (Guomindang) in overthrow of Imperial system.

1919 After Versailles Treaty deals with aftermath of the First World War, mass demonstrations of 4 May Movement objecting to Japan being granted Germany's Chinese colonies.

1921 Chinese Communist Party (CPC) formed in Shanghai; co-operates with Guomindang.

1927 Guomindang under Chiang Kai-shek's leadership attacks CPC, massacring its members and supporters in Shanghai and Canton. Mao Zedong urges CPC to form army, withdraw to countryside and create Red Bases. Civil war between Guomindang government and CPC begins.

1930–4 Chiang Kai-shek begins encirclement of Red Bases; 1934 communists begin Long March to escape from south and eventually regroup in Yanan in north China from 1935.

1932	After further Japanese aggression, communists declare war while Guomindang agrees to peace with Japan.
1937	Japan takes Beijing, while Chiang is now forced by his own generals to join with CPC in united front against them. Japan launches all-out offensive in east China.
1945	Japan defeated; USA destroys Hiroshima and Nagasaki with nuclear bombs.
1946	Civil war again breaks out between CPC and Guomindang. Liberated (CPC-controlled) areas implement land reforms.
1949	Guomindang defeated, Chiang Kai-shek and supporters flee to Taiwan and establish government in exile, maintaining hostilities with US support.
	1 October Mao Zedong declares Liberation and establishment of People's Republic of China in Beijing.
1950	Sino-Soviet Friendship Treaty.
	Rural reforms across entire country, including land transfers from landlords and rich peasants.
	Korean War starts; Chinese volunteers go to support North Korea.
1953–7	First Five-Year Plan, with large-scale support from the USSR for heavy industrialisation.
1958–60	Great Leap Forward movement encourages small-scale and rural development; coincides with formation of huge rural collectives (People's Communes); bad weather and poor incentives in communes lead to failure of harvests and unprecedented famine, with between 13 and 26 million deaths.
1959	Armed revolt in Tibet suppressed by Chinese army, and Dalai Lama leaves for India and exile.
1960	Sino–Soviet relations collapse and USSR pulls out of aid projects.
1962	In border clashes China defeats India on disputed boundaries imposed during British rule in India.
1964	China explodes first atom bomb in test.
1966	Cultural Revolution initiated, followed by three years of economic and social disruption, with schools and colleges closed, Red Guard groups formed.
1969	Border battles between USSR and China in northeast.
1971	China admitted to the United Nations.
1972	President Nixon visits Beijing and joint communiqué issued.
1976	Premier Zhou Enlai dies in January; he had been an important force in constraining the influence of the 'ultra-leftists'. In September, Party Chairman Mao Zedong dies, leaving the way open for the 'ultra-left' leaders, headed by the 'Gang of Four' (which include his widow Jiang Qing) to be arrested.

1979	After a crucial party meeting the previous December, which affirmed Deng Xiaoping's leadership, a series of major economic shifts are introduced which lead eventually to the disbanding of the communes and collective agriculture, and to the reduction of the state's role in planning in favour of a more market-oriented economy.
	Serious fighting with Vietnam ostensibly in Chinese protest at border violations, but really to pressure Vietnam after it occupies Kampuchea.
1981	Trial of the 'Gang of Four' and other 'ultra-leftists' ends with suspended death sentences for some and long prison sentences for others.
1982	The first census since 1964 shows there are just over one billion people in China.
1984	Britain agrees to the return of Hong Kong in 1997.
1986	Queen Elizabeth visits China, first British monarch to do so.
1986–7	Widespread student demonstrations revive the short-lived 'Democracy Movement' of the late 1970s, but are unsuccessful and bring a clampdown on freedom of expression. Hu Yaobang is demoted and removed from post of Party General Secretary; Zhao Ziyang replaces him.
1987	Autumn: serious political unrest in Tibet (and again in March and November 1988). Dalai Lama and Beijing seem willing to engage in talks on future status of the region.
1989	In May, mass demonstrations initiated by students in Beijing and other cities are given widespread support by the people. After martial law is declared in Beijing, troops massacre hundreds of people on and after 4 June. Party conservatives take control, and Zhao Ziyang is accused of supporting the demonstrators, who are accused of counter-revolution. Arrests of thousands of protestors follow, and executions begin.

Chapter one

Introduction: a basic guide to developments from 1949 to 1989

Terry Cannon and Alan Jenkins

In September 1976, Mao Zedong (Mao Tse-Tung), one of this century's major political leaders, died in Beijing (Peking), the capital of the world's most populous country. The significance of the subsequent immense changes in economic and social policy, and their consequences for the country's geography, can only be understood in the context of the events of the previous years. During the period of Mao's leadership China's communists emphasised rural collectivisation, central planning plus state control and ownership of industry, and limited contact with the outside world.

These policies became identified as essential components of socialism in the country. China's post-Mao leadership, dominated by Deng Xiaoping (Teng Hsiao-Ping), emphasised economic growth, the role of the market, family farming, much more foreign trade, and investment from the West and Japan. It claimed that these very different policies – the so-called economic reforms – were necessary to overcome the problems which Maoism created, and that they too are socialist.

Our book is focused on this fundamental shift in direction, and its impact on the geography of the country. The introduction helps to explain what is entailed in the policy changes and the political and economic justifications for them. For those readers with little knowledge of China, it is an important chapter for helping to make sense of the book. It tries to explain the major policies of the communist-led government which came to power in 1949 after years of civil war and Japanese invasion, and the major shifts in policy since then.

PARTY AND GOVERNMENT AS FACTORS IN CHINA'S DEVELOPMENT

In China the role of the state, and the Communist Party of China (CPC) which dominates it, has been crucial in determining what economic and environmental changes take place. It is difficult to understand the country's geography without knowing something of the role of the state since the

CPC came to power, and especially of the sharp contrasts between those policies in operation up to 1976, and those of the economic reform period which has followed with the 1979 rise to power of Deng Xiaoping and his leadership. While the bulk of the book emphasises the geographical situation since 1979, here in the Introduction we try to explain the main issues in the policy developments since 1949 and their contrasts to the post-1979 reforms. (A brief summary of the major policies is given on pp. 24–7.)

Conflict and disagreement over policy

The dominance of a single party in China can give the impression that the state acts in uniform and harmonious agreement upon unanimously-agreed policies. This is far from the case, though disagreement is more convoluted and less visible than in multi-party states, where differences of opinion are public and between obvious groupings. In China, argument goes on within the CPC, and divisions can be so intense that leaders who lose power struggles are denounced as counter-revolutionaries, and *coups d'état* have been used to remove opponents.

It is much more difficult to display popular discontent about the state's behaviour, though not impossible. The second half of Deng's decade has been marked by many very significant signs of dissaffection by important sections of the people. Widespread urban opposition to the negative impacts of the economic reforms grew in the latter half of the 1980s, and there was a resurgence of demands for more democracy. The events of Spring 1989, involving mass popular protests in many cities, and their brutal suppression by Deng Xiaoping and his supporters (led by the Prime Minister, Li Peng), make it unclear what is going to happen to economic policy and hence to their geographic impact.

Our original idea for this chapter, before the massacres, was to summarise the impact of a decade of influence of Deng's leadership. The tragedy of 4 June 1989 and its aftermath marks the end of that decade in a manner which highlights the conflicts that arose under the reform policies. It also signifies the beginning of a process of change, and as we write the Introduction it is unclear to what extent the post-Mao policies will remain intact.

Geography and economic policy

This book's central theme is the nature and extent of the changes since Mao's death. In the following chapters, various spatial and environmental issues are investigated. These include the physical environment and how, given its fragility, it is being managed; the impact of historical influences, as in the relationship between the previously more-developed coast and less-developed interior; how the new rural policies have changed China's landscape; how the new economic policies and pressure on couples to have only one child are affecting the relationship between an increasing population and limited fertile land; how regional differences are changing under the impact of rapid industrialisation and slow transport development; the extent that the surplus population released from agriculture is contributing to urbanisation; the impact on China's geography of the increased trade and investment from the West and Japan, and whether it is leading to increased inequality between the developed coast and the less-developed interior. Are regional contrasts and inequalities increasing throughout China as a result of the new policies? These are some of the spatial and environmental themes investigated.

The book's central argument is that China's geography has significantly changed as a result of the policies of the post-Mao leadership. This introduction emphasises the period 1949 to 1976 because to understand the changes a basic understanding of 'Maoist' China is needed. We then more briefly consider the main directions in economic policy after Deng's ascendancy in 1979. Our aim here is to enable readers who know little about China to understand the general development policies resulting in the geographic changes set out in later chapters. The Introduction and later chapters also reveal the impact of China's natural conditions and human geography on development opportunities.

Background to 1949

In 1949 the Communist Party of China (CPC) became the effective government of the country after victory in the civil war against the Guomindang (Nationalist Party). The CPC faced the immediate problem of rebuilding a society and economy that since the 1920s had suffered civil war, Japanese invasion and then again civil war. There were also more fundamental problems, rooted in the inherited exploitative system fostered under the Emperors. Whatever the achievements under that Imperial past, the economy had failed to make the technological and organisational breakthrough to 'modernity' achieved by the West and Japan.

The CPC achieved its victory on the basis of mass support for its struggle against the Japanese invasion, and the land reform implemented in the areas it controlled in the previous decade. But it inherited fundamental problems, and there were few guidelines for resolving them. China's population was increasing, the fertile land was limited. The modern sector of the economy was restricted to the port cities, largely fashioned by Western imperialism, and the Japanese-developed industry in the north-east (Manchuria).

In 1949, Mao Zedong, Chairman of the CPC, proclaimed that China had 'stood up' and that the party would build a 'prosperous and flourishing country'. Forty years later this remains to be achieved, and argument goes on as to whether Mao's own methods have delayed reaching this goal or made it more possible. Certainly the policies which followed him have negated his ideas as to what are proper socialist policies. However, the growth in prosperity evident for many people in the 1980s under Deng's leadership has led to its own problems. These are turning out to be as difficult to resolve as those for which Mao's policies are indicted.

FROM LIBERATION TO THE DEATH OF MAO

The Communist Party can be seen as but one group of Chinese reformers of the last century or so who wanted China to withstand outside aggression and achieve modernity. Although there are disputes about how the party interpreted socialism and communism, and the extent to which it departed from Marxism or Soviet thinking and practice, there can be no doubt that the CPC believed that economic growth and modernity was to be achieved through 'building socialism'.

In practice this meant that the Communist Party of China controlled political power and made the central decisions on economic and social policy. The Soviet Union initially became the model for the organisation of government and society. In the post-1949 economy there were three key areas which can be identified as its declared socialist elements:

1 in rural affairs, the collectivised ownership of land and other means of production, and a state-directed plan for the procurement of foodstuffs and industrial crops;
2 centralised control over the accumulation and reinvestment of capital, in combination with state ownership of major industries and financial institutions (and, in effect, local state ownership of smaller-scale enterprises);
3 restrictions on the impact of foreign capital and external economic factors while pursuing 'self-reliance'.

We look at each of these elements and their implementation in the period 1949 to 1976. All of them have been affected drastically by the policy changes pursued by the post-Mao leadership under the direction of Deng Xiaoping, and later we describe how far (at least, until mid-1989) they were being 'reformed'. In many senses it is clear they have been reversed rather than reformed.

Rural China: land and collectivisation

After 1949 the new government completed the land reform already begun in areas it had controlled in the civil war. This massive redistribution of land away from landlords and the richest peasants was welcomed by the

4

majority, who were poorer peasants. But for the leadership, this greater equalisation of land-holdings was a temporary objective. Private family farm-holdings were not regarded as socialist policy, and in any case many in the party argued that the allocation of land would not remain equitable for long. New divisions, exploitation and uneven land ownership quickly showed signs of re-emerging.

Perhaps there were other motives that caused the party to sweep away this newly created pattern of family farming. The land reform did not generate much capital for family or state investment. From the mid-1950s the CPC enforced the collectivisation of rural land, equipment and animals. The new collectives were given production targets to supply the state with part of their output of grain and industrial crops (e.g. cotton) at state-defined prices. This tied them into the national plan through the local administrative structure. Targets and prices were transmitted downward through the provincial structure to the individual commune and thence to the team.

Collectives offered the means of pooling labour-power and capital to transform rural China. Communal labour facilitated the improvement of land, the construction of dams to conserve limited rainfall, and irrigation systems. It enabled surplus to be generated to pay for schools and health centres, often the first ever available to ordinary people in many rural areas.

From the perspective of the central government, collectivisation may have offered a pragmatic means of getting secure and cheap grain and other crops and selling them at state-defined prices. Whatever the reasons, the CPC transformed rural China into larger and larger collective units. In 1958 the co-operatives set up after 1954 were merged into the People's Communes. The rhetoric of the time told us that this was essentially voluntary and popular. However the evidence now is that this move was forced through rapidly by Mao Zedong and others in the face of opposition from within the party and with little support from the peasants.

The introduction of the People's Communes was linked with the Great Leap Forward (GLF) of 1958–60, a foolhardy attempt to achieve full communism virtually overnight. 'Free' collective kitchens, communal childcare and universal industrialisation were promoted by Mao Zedong throughout the countryside. The idea was that mass enthusiasm and the new larger collective organisation would produce rapid economic growth.

The reality was that China plunged into disaster. The years 1959–61 witnessed a famine across large parts of China. Outsiders have only in recent years recognised the extent of that disaster, and the Chinese authorities are still reluctant to discuss it openly. At the time, a few Western observers, using interviews with refugees in Hong Kong and other sources, correctly told of the horror, but their accounts were not widely believed. The evidence we now have, including that from the 1982 Chinese census, reveals the worst disaster in the world this century, far outstripping the recent Ethiopian famines in absolute numbers. It appears that over the three years 1959–61 there were between 14 million and 26 million excess deaths attributable to hunger and associated illness.

A revised form of commune, which survived the criticisms of the GLF and was intended to extract China from the famine disaster, became the basic building block of rural economy and government until the post-Mao changes. On average a commune had about 15,000 members, but numbers varied from around 80,000 in the densely settled fertile areas of eastern China to 8,000 or less in sparsely populated mountainous or arid areas.

The commune was both a production organisation and a unit of government. It was divided into production brigades, which were in turn subdivided into work teams. Generally the brigade accorded with the 'natural' village, so to an extent the peasantry lived in a traditional setting. But the organisation of rural life was completely transformed from its former pattern. Generally families had only a small private plot on which they might grow what vegetables they wished and perhaps keep a pig or poultry. Often there were restrictions on the sale of this produce in the limited private markets. (Note that most policies varied in time and space; we are here describing the general pattern for the period 1958 to 1976.)

Peasant life was largely organised around the collective, in which the basic unit was the production team. This organised labour and capital, and distributed rewards. The key to the team's development role was the work-point payments system. This was like a dividend: work points represented a claim on the collective output of the team, supposedly proportionate to individual effort. Rural electrification, small-scale rural industries (e.g. brick works), irrigation schemes, clinics and schools could be built by mobilising labour which did not have to be immediately paid in cash. The work would be done in return for work points, since this increased people's eventual claim to the produce of the team.

Periodically, the team would distribute produce and income to the families on the basis of the work points the family had earned and the income the team had gained through selling its produce on the state-controlled markets. Whereas in other Third World states 'surplus' rural labour migrated to the cities, this did not occur in China (but see Chapter eight).

The collective mobilisation of labour provided a means of utilising this 'surplus' labour which increased production and investment. Many foreign observers (including the present authors) were attracted by this development model, which seemed to offer development while apparently avoiding those problems of landlessness and poor migrant shanty towns that characterised other Third World states. This view is apparent in the work of the geographer Keith Buchanan in *The Transformation of the Chinese Earth* (1970), a moving and sympathetic portrayal of the geography of Maoist China.

Such portrayals usually failed to discuss the tight controls on migration, especially to the cities. Control was achieved by allocating key commodities in the cities such as grain and pork through rationing, and by requiring residency permits for urban areas. The post-Mao leadership has criticised the communes for reducing the incentives for individuals so that output was too low and labour wasted. As we will see, the abandonment of the commune system in favour of individual family farming has revealed that there is considerable surplus of labour as required for agriculture, much of which has moved to small towns and cities.

Geographically, one of the most important aspects of Maoist policies was the emphasis on regional and local self-sufficiency. A common slogan was 'Take grain as the key link, promote all-round development'. This signalled the regime's awareness that adequate grain output (especially wheat and rice) was vital to preventing both famine and urban unrest. It also indicated the importance of local self-reliance, the idea that local areas should pull themselves upwards out of poverty. Maoist policy created a particular kind of organisation and a particular geography in rural China; recent policies have been transforming that geography.

Dazhai (Tachai) production brigade in Shanxi province became the model of Maoist self-reliance for other rural units. Though previously very poor, in an area with very difficult physical conditions (an arid region of eroded gully-land in the loess plateau), it had apparently achieved greater communal prosperity through collective effort.

If the team and the commune were the organisational units of rural life, their activities were shaped by the decisions of central (and provincial) governments. Decisions on priorities between investment in agriculture, industry, transport and other infrastructure, welfare, and the location of investments were made by the bureaucratic organisation of government and party. In agriculture, decisions on the priorities between different crops were largely made by central and then provincial governments setting output targets and determining prices. In theory, the lower levels of the hierarchy could communicate their views upwards.

Chinese (and most Western sources) now insist that in practice the system was essentially a top-down model, and that local officials were under immense pressure to fulfil central directives which were often insensitive to local conditions. Despite Maoist emphasis on rural development, most of the effort was expected to come from collectivised resources and local self-reliance: there was little direct government investment in agriculture.

Central control of capital: industry and the cities

The Communist Party in 1949 had scant experience in managing either cities or large-scale industrial enterprises. The revolution had been fought largely in rural areas, based among peasants rather than workers, and most of the leaders had little idea how to run the cities. For a period, existing industrialists were permitted to operate more or less on their own. Then, from the mid-1950s, major changes in ownership were begun, involving nationalisation.

There were few models of industrialisation available to a new socialist government, and it is understandable that the Soviet Union became China's mentor. Not only did the USSR appear to have been very successful industrially, it was willing to lend capital and provide technical assistance. In any case, communist wisdom had it that socialism was virtually to be equated with large-scale enterprises and the promotion of heavy industry, a model which the Soviets had pioneered.

In 1953, China launched the First Five Year Plan (FYP), modelled on Soviet lines, with state control of most urban industry. Profits from industrial production (incorporating the profits made from the cheap purchases from the agricultural sector) were directed into what were seen as key sectors that would make China prosperous and strong. Iron, steel and textile industries were among those particularly developed.

Chinese economic planners shared Soviet thinking that (Western) capitalism before 1949 had produced in China unequal regional development. The first FYP involved the construction of new inland industrial bases in cities such as Wuhan, Xian (Sian), Lanzhou and Chengdu. However, as early as 1956 Mao was signalling the 'pragmatic' advantages of development in the established coastal region, betraying a belief in the virtues of regional inequality (see Chapter two).

The close relations between China and the Soviet Union, and the basic similarity of models of economic development, were both short-lived. A variety of issues (including Mao's questioning of the Soviet model) led to conflict between the two communist giants (see Chapter eleven). From 1958, a section of the Chinese leadership under Mao's domination was able to promote an entirely new development strategy, the Great Leap Forward (GLF). It ended in 1960, a disastrous failure, and so too did the special relationship with the USSR. Soviet aid was cut off, and its technicians abruptly withdrew. Mao was forced into the background of economic policy as a result of the failure of the GLF and the associated famine tragedy.

However, the GLF did produce the valuable new concept of rural industrialistion. Ironically, this provided a legacy without which today's dynamic rural enterprises would probably not exist. But in the urban sector little was altered and the concentration on heavy and large-scale industry continued as before in the first FYP.

This emphasis on predominantly urban and large enterprises was accompanied by a respectable rate of industrial growth, even during most of the Cultural Revolution. But capital was used wastefully, being allocated by the state to be used by enterprises without any charge (such as interest payments). This meant that more capital was needed compared with the goods produced for peoples' consumption using that capital.

Living standards were therefore relatively stagnant, a victim of the high savings ratio (i.e. the high proportion of funds used for productive investment compared with the amount left available for peoples' consumption). Consumer goods were in any case not given priority, and shortages of many ordinary products, as well as 'luxuries' like bicycles, radios or varied clothing materials, were normal.

The industry fetish also meant a low level of investment in urban infrastructure and welfare, again affecting living standards through the inadequacy of housing, public transport, sewage networks and electricity supply. Limited investment in basic infrastructure affected the productivity of industry itself: too little capital was directed to power generation and transport (see Chapter seven).

The Chinese media now emphasise the shortcomings of the pre-1979 commandist industrial planning system. Yet we should remember that a respectable rate of industrial growth did occur. The modern sector of the economy was increased both in geographical extent and in terms of absolute production.

As with rural China, the industrial-planning system was criticised in the 1980s for its highly centralised, bureaucratic nature. Factories could be constrained by shortages of components or raw materials, owing to the intricacies of the quota target system which determined the production of the state-run sector. At the other extreme, the quota might be reached but the product be of unusable quality, or no longer needed. This strategy was perhaps more suited to the initiation of industrial growth but less appropriate for long-term development. Criticism has also been directed at the system's reliance on production (and prices) being determined by bureaucratic decision rather than by the market.

Foreign trade and investment

Finally in this discussion of industrial China we should mention the changing attitude to foreign trade, and the role of foreign investment and technology. As we have seen, in the 1950s China did use Soviet technology, capital, and expertise to help modernise its industrial sector. However, from 1960 China largely had to go it alone, and until the 1980s foreign trade was a very small component of the economy.

In the 1960s and 1970s, Maoist development policy emphasised the benefits to China of 'self-reliance'. A statement from the 1960s encapsulates that thinking: 'A country should manufacture by itself all the products it needs whenever and wherever possible ... [self-reliance] also means that a country should carry on its general economic construction on the basis of its own human, material and financial resources' (cited in Howe 1978: 135). Such statements of principle may have been shaped by pragmatic necessity, since trade and access to capital and modern technology were limited. China was in conflict with either or both the Soviet Union and the USA for most of this period. China's post-Mao leadership has taken a very different view of the role of foreign trade and technology (see Chapter nine).

CONFLICTS AND CHOICES FROM 1979 TO 1989

We have seen the impact in the period 1949–76 of three elements considered crucial to socialist development in China: the collectivisation of land and state-directed agricultural planning; central control over the accumulation and re-investment of capital, coupled with state ownership of major industries; and a policy of 'self-reliance' and the exclusion of foreign capital. However our description has only touched on the policy conflicts and choices in China's post-1949 development strategy. To explain what happens after Mao's death, these issues need now to become central.

As is clear from the above discussion, communist ideology was capable of generating conflicting strategies in China's post-1949 development. The problems facing the Chinese leaders in 1949 were horrendous, including a large and poor population, limited fertile land, a vast territory with immense transport difficulties, and limited capital. While much has changed in the forty years since then, these and other problems remain. In 1949 there was broad agreement in the party that socialism, involving the three key components discussed above, was the way to deal with them. In the 1980s many of the self-same leaders presided over policies which admit to the limitations of those three aspects of socialism.

Mao's death and the rise of Deng

Soon after Mao Zedong's death in 1976, there was, in effect, a *coup d'état* which overthrew a small group of party leaders whose power depended on Mao's patronage. A section of the party, combined with groups in the security forces, installed a new CPC Chairman, Hua Guofeng. A number of the 'ultra-leftist' political leaders associated with the policies of the Cultural Revolution of 1966–1976 were imprisoned. Chief among these was the 'Gang of Four', which included Mao's widow, Jiang Qing. In 1980 they were tried and convicted of undermining the revolution and using illegal means to defeat their opponents (including imprisonment, torture and persecution to death).

The change-over of power was peaceful but not uncontested, and there was conflict over the role of Hua Guofeng and others, who supported a continuation of 'Maoist' economic policies. The two years following Mao's death were marked by intense political in-fighting, and exponents of contradictory economic policies tried to gain the upper hand. By increasingly blaming Maoist strategy for the country's economic problems, sections of the CPC which favoured economic 'reforms' were brought together by the so-called pragmatists.

By 1978 a party grouping around Deng Xiaoping had coalesced, established a clear power-base and a new direction for economic development. At an important CPC meeting in December that year, Deng was victorious in establishing his policies for economic reform as the official line. The Four Modernisations became the watchword, and politics and class struggle were subordinated to economic progress.

It is from December 1978 that the new direction in development policies which are still transforming China's geography can clearly be identified. At that CPC meeting (the Third Plenary of the Eleventh Central Committee), the party publicly repudiated the Maoist slogan 'take class struggle as the key link'. Ideological correctness, whatever the price for people's lower living standards, was replaced by a 'pragmatic' approach to problems: the party should be 'acting according to objective economic laws'.

The Chinese media used a phrase endorsed by Deng that 'Practice [experience] is the sole criterion of truth'. Such slogans have to be taken seriously as an important form of political discourse in China. It is this major shift in policies which decided that we use the year 1979 as the beginning of the reform period and of Deng Xiaoping's decade.

In Chinese politics, especially before 1979, policy conflicts were often discussed in terms of the supposed opposites 'red' and 'expert'. Many of the older leaders, veterans of the revolution (and some, like Deng, in their eighties during the reform decade), would regard their credentials for ruling China as 'red' rather than 'expert'. A 'red' position would emphasise socialist revolutionary goals, and see problems as being solved by having a 'correct' political attitude rather than the particular expertise which is assumed necessary in bourgeois society. An 'expert' position would emphasise knowledge of a particular subject as the key to problem-solving. But Deng, a member of the party since 1924, had held key positions in it and, was clearly a proponent of expertness and a fierce critic of many of the 'Maoist' 'revolutionary' policies. He is reported to have said 'If a cat catches mice, what does it matter if it's black or white'. In the context of China, this means, 'Don't ask whether a policy is socialist, ask whether it works'.

This begs the question as to what criteria should be used to know if it is working. Success can be measured differently by people (or interest groups) who have contrasting attitudes. The liberation of women, for example, can be argued to be an essential component of socialist political change (and there is evidence of significant progress in Mao's time). But a pragmatic, 'expert', view might see such policies as economically wasteful.

An important Cultural Revolution policy was the introduction of basic medical training for about two million rural people (the 'barefoot doctors'), with the aim that health care be spread from the urban areas in which it was concentrated. The 'experts' – doctors in the cities and towns – would probably not have gone out to do the training without the politics of the campaign.

Although most Chinese people do not know the truth about the massacres of June 1989, many realise from the problems associated with Deng's reform programme that pragmatism and 'expertness' are no guarantee of economic success. His supporters might argue that contention over the reforms meant they had to be compromised and left incomplete, preventing them from being properly implemented. But the unwillingness of the CPC to reform itself and reduce its power, and so bring about political as well as economic change has embittered significant sections of the people. After martial law had been declared but before the army massacre in Beijing, students had put up posters saying 'What does it matter the colour of the cat so long as it resigns'. That many died making their peaceful protest means that Deng will be remembered very differently by at least some of the population.

THE FOUR MODERNISATIONS AND ECONOMIC REFORM

Modernisation and economic growth have been the main political goals of the past decade. The leadership has called on the people to achieve the Four Modernisations (agriculture, industry, education, and science and defence). This slogan is revived from a speech by Premier Zhou Enlai in 1964. The attitude it embodies was curtailed in the Cultural Revolution (1966–76), but at the 1978 Third Plenary of the CPC the policy shifted and emphasised the achievment of the 'Four Modernisations' through 'readjustment, reform, consolidation and improvement':

> Readjustment meant shifting priorities from industry to agriculture, from heavy industry to light industry, and focussing attention on bottleneck sectors. Reform came to mean moving away from the Stalinist-style command economy and embarking on the types of economic reform already implemented in Eastern Europe. ... Consolidation and improvement meant essentially trying to upgrade the level of management and efficiency.
>
> (Lockett, 1986: 35)

The details of these policy developments are taken up in the following chapters. Here we will emphasise key aspects of development policy under

Deng's leadership in the reform decade, in particular indicating how they have altered those central features of China's policies which previously were considered vital components of its socialist principles: the collectivisation of land and state-directed agricultural planning; central control over capital with state ownership of industry; and 'self-reliance'.

Rural China: family farming and the market

In the countryside, the changes from Mao's policies came early and took effect quickly. The impact of the reforms have been extensive and fundamental, and the collectivised landscape is now gone. The new policies for rural China (analysed in detail in Chapter six) are rooted in the 'responsibility system', which represents a return to family farming.

The communes were criticised on the basis that the work-point system meant that personal incomes were not sufficiently sensitive to the amount of individual effort put in, so that allegedly those who did not work so hard were rewarded with not much less than those who did. There was a basic ration of grain allocated to all before work points were assessed. So the diagnosis of the rural problem (low incomes and slow growth) was of a failure of incentives.

Under the reforms, families have been allocated land on leases (of up to thirty years), in return for which they contract to sell to the state, at a fixed low price, an agreed quota of grain or industrial crops. The remainder of their output is used or sold for consumption, and any surplus can be sold (at higher prices) to the state or on the private markets which grew rapidly during the 1980s. The new rural strategy was based on the assumption that increased productivity would occur through 'releasing' individual and family enthusiasm and effort in the responsibility system.

Alongside this private farming, a tremendous growth in rural business initiative has occurred, in commercial and manufacturing as well as agricultural enterprises. Some of the older collective workshops and sidelines developed under the commune system (like piggeries, fishponds and small factories) have been dispersed, while others have been leased out to groups or private individuals. More crucially, completely new enterprises have been springing up entirely in private hands.

Such economic growth in villages and small towns is central to the development strategy of post-1979 China, given Chinese estimates that about 30 per cent of rural population is deemed to be surplus to the needs of agriculture. One of the issues explored in Chapter eight is the nature and extent of the urbanisation permitted and even encouraged during the last decade.

Under the rural reforms, farmers have been producing more, and the leadership considers this to be a response to the incentives inherent in the responsibility system. Extra income can be earned by individuals and families once the state quota and personal consumption needs have been met: peasants are free to dispose of any surplus to earn cash. Near the larger towns and cities the policies have also led to the development of rural industry, and the higher personal incomes have produced a boom in new rural housing. But these developments are eating away at China's limited arable land and posing severe problems of pollution and environmental management (see Chapter ten).

This also raises the whole issue of population growth, and the reform leadership's conviction that the country is overpopulated. Increased production represents one way of solving China's problem of the relationship between people, land and production. Under Deng's leadership the country has also sought to limit population growth much more than in the previous decade, though policies such as the one-child family have been much more effective in urban China.

Does the rural prosperity apparent in some areas under the new system derive from the legacy of the facilities built with collective labour in the previous commune system? The building or maintenance of rural infrastructure, of water conservancy projects and schools and clinics is being neglected under the new system. Without such communal efforts the new prosperity may be short-lived and the current wealth earned at the expense of running down local facilities. There is now concern that rural output has already peaked. There is a dangerous combination of labour productivity which cannot be any further increased and the lack of any system in place to promote the regeneration of agricultural investment and infrastructure development.

In Maoist China rural policies emphasised self-reliance and collective effort. Where the policy worked it created an upward spiral of development involving social and welfare facilities as well as agricultural and local industrial growth. But it was not successful everywhere, and relied to some extent on the use of exemplary collectives whose experiences were crudely and inappropriately promoted throughout the country. Denigration of such models following a policy change is very typical in China. Dazhai is no longer seen as the answer to rural poverty, but it also has to be discredited because of its symbolic role. This model Dazhai production brigade has been attacked since 1979, with allegations that it secretly obtained state support to facilitate its growth.

In contrast with Mao's time, post-1979 policies emphasise that peasants should decide what to produce, often on the basis of what maximises the advantages of natural conditions and location. The market has increasingly replaced government and party as a mechanism for stimulating what is produced locally. Regional self-sufficiency (and the Maoist emphasis on grain) is criticised for creating a situation of shared poverty rather than the prefered differential growth: 'Let some localities in the country . . . prosper first . . . for they will set an example to others' (*Beijing Review* 19 January

1981). Now market forces are to be used to encourage regional specialisation, and regional inequalities are accepted as necessary to building 'socialist modernisation'.

Urban China: decline of the state's role in the economy

The slogan 'readjustment and reform' reveals two different policy directions. Readjustment suggests that the pre-existing system was basically fine and just needed adjusting. Reform suggests that something more fundamental was required. During Deng's decade rural China had certainly been reformed: supporters of the commune system might say that a more accurate term would be counter-revolution.

Changes to urban China have taken longer and proved more difficult to implement. It has been difficult for some party leaders to accept the ideology of a reduction in the state's role in directing the industrial economy. Large-scale, state-owned enterprises employ many people and represent much of the state's capital investment. Changes in how these should be owned or managed, and in who decides what is to be produced and sold at what price created much controversy.

Significant changes to the urban industrial sector did not take effect until after 1984. In October that year, the Twelfth Central Committee of the CPC stated that 'defects in the urban economic sector . . . seriously hinder the expansion of the forces of production'. This meeting set out the outlines for reform. Its key features included a cutting back of party control over the 'rigid economic structure', with the government (both nationally and locally) withdrawing from management and taking a more regulatory role. There was still to be an emphasis on state planning, but mandatory planning was only to cover key commodities like steel, and large-scale infrastructure such as railway construction. Urban enterprises were to become more independent and to be 'responsible for [their] own profit and loss', with managers having more power. The price system should be reformed and prices reflect 'economic laws' (e.g. supply and demand).

Editorials in *The People's Daily*, the main party newspaper, said that the prices of goods would be determined by their production costs, that the law of value would become 'a special feature of Chinese socialism'. Wages should reflect productivity; egalitarianism and workers' belief in a secure job for life regardless of effort (the 'iron rice bowl') were attacked. Regional industrial specialisation was to be encouraged, and this has been linked with a consistent policy to promote the more rapid development of the coastal region, which is seen to have an initial advantage and lower production costs (see Chapter two). This has not been accepted without dispute by other parts of the country. This and other problems of the urban reforms are discussed in the final section of this chapter.

Foreign trade and capital: the 'open door' policy

Foreign investment and an expansion in trade have been central to Deng's modernisation strategy, and self-reliance is a policy of the past. To promote China's new role in the international economy, very early in the reform period the leadership declared an 'open door' policy. This was designed to improve business confidence among potential traders and investors, and began the process in which reduced tax liabilities and other concessions have been used to entice investment from outside sources.

The rationale for breaking with the previously defined socialist policy was that if the Four Modernisations were to succeed, then new technology had to be brought into the country. Initially, the idea was to focus developments in the four Special Economic Zones (SEZ) on the south coast, partly to isolate the decadence of capitalist influences. But, in fact, over the reform decade many more coastal cities and areas have been opened up to encourage foreign investment, and there is acceptance of such involvement virtually anywhere (see Chapter nine).

The decision to use foreign technology in much larger amounts than before has serious 'knock-on' consequences throughout the economy. Some could be introduced by foreign firms locating in the SEZs and being taken over by Chinese at the end of the contract. But this route has not been very successful, and the SEZ policy has been criticised for this shortcoming. To buy expensive imports of equipment involves increasing China's exports, and this entails using scarce resources which can also be of use within the country. The sort of costs-versus-benefits analysis that is entailed in this cannot be simply economic. Considerable political opposition has arisen, for example, over the massive exports of oil to Japan used to pay for technology from that country; given the severe energy shortages in China it is fairly easy to guess the arguments.

It has been difficult for the government, given the increased autonomy of enterprises and local authorities, to control the level of imports. While the manufacture of goods for export has increased enormously in the 1980s (as have invisible trade earnings such as tourism), the total value of all exports has often failed to keep up with imports, resulting in very large balance of payments deficits in some years.

Instead of earning money through exports in order to buy technology, China has also broken with one of the strictest aspects of Maoist policy and borrowed money from foreign banks and governments. Moreover, in the early 1980s it confirmed its new role in the international economy by joining the World Bank and other institutions previously regarded as tools of the imperialist system.

Since much of the new foreign investment, and a good deal of the increase in export industries, are located in the coastal region, the 'open door' policy has effectively reinforced the state's preference for the more rapid growth of that region, and this has added to the inter-region tension in the country (see Chapter two).

A DECADE OF REFORM AND DENG'S LEGACY: HOPE OR DESPAIR?

It was easy for the post-1979 leadership to set out a critique of Maoist policies but far harder for it to implement new ones to replace them. As the various reforms in the Soviet bloc have demonstrated, it is difficult to find a structure that both retains elements of central planning (so as to pursue socialist goals which it is felt will be smothered without it), and to introduce market forces to improve economic performance, higher income and growth.

China's reforms have gone through two phases, with 1985–6 marking the turning point between a period of apparent success in terms of the reformers' aims (1979–85), and one of apprehension or despondency among both the CPC leaders and significant sections of the populace (1986–9).

The earlier period saw the rural reforms take hold throughout the country, with an associated rise in agricultural output and peasants' incomes, and the beginnings of the new commercialism of the countryside directed to small towns and cities. The reforms seemed popular with most sections of society, though early attempts by some to encourage political democracy (in the Democracy Movement of 1978–9) were repressed. Many urban dwellers felt the improvement of more choice of foodstuffs, and the increased output of light industry meant there were more consumer goods available.

Policy conflicts and the reforms

There is no single event which clearly marks the downturn in the fortune of the reforms; a combination of factors has been involved. Some seem inherent in the reforms themselves; others are products of political obstruction by some leaders of the proper implementation of the reforms, leading only to their partial or imperfect implementation. In addition, the fact that political reform has not accompanied changes in economic policy has made co-operation with the government much less attractive to some crucial groups of people.

The party (and its reforms) has become discredited as local and national leaders (and their relatives) cashed in on the commercial opportunities arising under the new policies, in which they had a head start. The new Democracy Movement of 1986 and 1989 was not just inspired by the contacts with the West which came with the 'open door' policy. It wanted political changes to restore certain aspects of the egalitarian elements of the past which were seen as having been eroded by the reform policies, though retaining the higher levels of wealth generated by them.

Since the mid-1980s the leadership has shown significant divisions between those who favoured retaining a more central role for the party in directing the economy (the 'conservatives'), and the 'liberals' or reformers (led by Deng Xiaoping and Zhao Ziyang), who have wanted to bring in much more of a market-led system.

The difficulties of the reforms result partly from the compromises forced by this conflict. But other problems are a direct result of the reforms, including the 'overheating' of the economy (leading to severe imbalances between the supply and demand for some basic goods), the mismatch between industrial growth and the lagging transport and power sectors (reducing efficiency and increasing the extent of power cuts), and the differentials in prices between the state and private sectors.

These have contributed to the severe inflation in prices of food and other daily necessities, which in the second half of the 1980s has rarely been below 20 per cent per annum, and which has, in several years, been 30 per cent or more. The problems outlined just now are also the basis of much of the corruption complained of by the people in the 1980s, which was a target of the 1989 protesters. While the reforms were in many ways popular, the party is not trusted to implement them fairly or with equality of opportunity.

After the 1989 massacres: changes in policy?

The events of May and June 1989, involving the violent crushing of popular demands that the party end the corruption which the reforms of the 1980s engendered, throw into doubt the continuation of the reform policies in their existing form.

The political instability which the reforms generated led to conflict in the party leadership over how to deal with it. After several weeks of confusion in May, supporters of Deng Xiaoping and Premier Li Peng in the army and party came together to crush the demonstrations and remove from the leadership those who would have granted some (probably very limited) political reforms. Zhao Ziyang, the General Secretary of the CPC and the man responsible for the implementation of the economic reforms (with Deng's support), was removed from office and accused of splitting the party. In his place Jiang Zemin has been installed.

In mid-1989, after the massacres, the 'conservative' wing of the CPC has been strengthened. It is not only firmly against political reform and democracy, but also includes elements who are opposed to the much more extensive role for market forces and commercialism which the reform leadership, especially Zhao Ziyang, had pushed.

Ironically, this group of leaders is also much more likely to try to deal with the issue of corruption and cleaning up the party, one of the demands of the mass demonstrations, because it is seen by them as a by-product of the over-commercialisation of the economy and reduction of central direction.

The 'two-track' pricing system of some products (especially raw materials and producer goods) led to the corrupt manipulation of the price differences (between fixed state prices and free market prices of the same goods) by those in leadership or management positions in party or enterprises. A strengthening of party direction of the economy is likely to try to deal with this, although the power of the Beijing leadership to control matters effectively in the provinces is more doubtful.

Deng Xiaoping, an opponent of political reform, maintained his position after martial law and the massacres. Have those events brought into question the continuation of his economic reforms? Did even he prefer the maintainance of party power to the continuation of the same type of economic change? A number of alternatives present themselves to us as we try to assess the short-term impact of the June events.

One thing seems certain: so long as he and the older party leaders are alive they will protect their position to enable them to pursue their own conception of socialism. This is no longer defined in terms of basic principles, but as whatever they themselves feel is appropriate at a given time, provided the CPC retains its power. They see themselves as the original revolutionaries who made socialism possible in the country for the benefit of all. This means that repression will take precedence over democracy, because any opposition is considered to be 'counter-revolutionary'.

Such a situation entails internal tensions that will make government difficult, with a populace unwilling and coerced. But need this necessarily affect the path of the economic reforms? Soon after his appointment, the new party leader, Jiang Zemin, stated that 'China's policy of economic reform and opening to the outside world would never change'. It could, he suggested, now be carried out even more effectively.

In the same vein, Deng spoke to army troops who had been involved in the Tienanmen massacre, saying: 'Generally, our basic proposals, ranging from a general strategy to policies, including reforms and opening up, are correct. If there is any inadequacy, then I should say our reforms and opening up have not proceeded adequately enough. . . . What is important is that we should never change China back into a closed country' (*Beijing Review* 10–16 July 1989).

The leadership also seems keen to reassure foreign groups that the country is still open for business. In spite of some trade sanctions immediately after the massacres, it is unlikely (given their past record with China and other repressive governments) that Western and Japanese investors and governments will be put off for very long. If they are, the delaying factor is likely to be concern about political stability affecting their investments rather than moral sensibilities.

Problems for the old and new leadership

Many of the basic questions about the running of the economy will remain to be dealt with, even if the post-massacre leadership makes party control its priority. What form of ownership does the CPC encourage now, and with what mix of state, collective and private ownership? What proportion of (state-controlled) capital is to be invested in agriculture, heavy industry or light industry? How much of society's 'surplus' is to be invested in new production capacity, and how much is set aside for collective or private consumption? How much does the party emphasise the goals of increased production and how much does it emphasise the goals of building a socialist society?

The restoration of family and private farming, and with it the expansion of the market for labour, raises the question of what is socialism in an essentially peasant society? Will gender issues ever be seen as equal to or more fundamental than those of class? What policy should the party adopt towards intellectuals, many of whom are now hostile to the regime but vital to modernisation? Can the country afford a nuclear weapons' strategy? If the government decides against it and instead directs that capital and intellectual resource into something like improved water control, what does it do if threatened by a nuclear power? Should China obtain technology from the outside world and continue to risk the problem of debt and the export drive?

The reactions to all of these issues will have an impact on China's geography. Some are more explicitly geographic, including the choice of investment in 'established', largely coastal industrial areas or in the poorly developed interior. Should agricultural investment be directed at the most fertile productive lands, the poorer hilly areas, or at strategically sensitive border areas? What policy should be adopted on the relationship between population growth and limited fertile land? Can China both 'develop' economically and conserve its fragile environment?

All these choices will continue to produce conflicts within the party. China's political system in the last forty years has been dependent on single, prominent leaders, especially Mao and Deng, who have become representatives of particular political outlooks. The traditional political culture of the country has been adapted by the CPC to its own needs.

The power-broking balance and patronage held by party leaders at all levels is at its most significant and unstable level at the top. As a result of investing so much importance in individual leaders the party is bound up in a potential crisis each time a major leader dies. The succession in the case of Deng Xiaoping is likely to be very difficult and painful, and is made all the more difficult by the role he played in the repression in 1989. It is likely that the army (or sections of it) will claim a much more prominent role in the management of both the economic policies of the country and the power struggles which will undoubtedly be a feature of the country as Deng and the other elderly veterans die.

In few countries of the world does a governing party and its leader have such power to determine the nature and direction of social and economic change. As will be seen in this book, this has also had its effect on the geography of the country: China's human geography and environment are more directly influenced by top-down policy than most of the world. Much of the rest of the book charts that situation over the last forty years, with particular emphasis on the period 1979 to 1989.

REFERENCES

Buchanan, K. (1970) *The Transformation of the Chinese Earth*, London: Bell.
Howe, C. (1978) *China's Economy: A Basic Guide*, London: Paul Elek.
Lockett, M. (1986) 'Economic growth and development' in D.S.G. Goodman *et al.*
    The China Challenge, London: Routledge.

UPDATE

The English-language journals *Beijing Review* and *China Reconstructs* (renamed *China Today* since January 1990) are useful sources of official Chinese statements. Many of the Western 'quality' newspapers such as *The New York Times*, the *Guardian*, *The Financial Times* and the *Independent* have good China correspondents.

The main academic journal on China, *The China Quarterly*, carries articles on recent developments, and a thorough chronicle of events, the 'Quarterly chronicle and documentation', with most useful excerpts from mainly Chinese sources. Two other journals, *Modern China* and *Australian Journal of Chinese Affairs*, are also useful, though less focused on recent events. The quarterly magazine *China Now* is a lively means of keeping abreast of recent changes, and often has issues on particular themes.

The book-review sections of all these periodicals inform about recently published books. *The China Quarterly* has a regular Books Received column which is very comprehensive on English-language books about China.

In February each year *The Far Eastern Economic Review* carries a survey on China, focusing on economic issues. Its regular weekly coverage usually has something of interest on China. The September issue of *Current History* annually is devoted to articles on recent developments in China. The annual Encyclopaedia Britannica *Book of the Year* provides a review of events in China for the previous year, also statistics on China's economy and society.

Articles on geography, and related areas of value like environment, economics and politics appear in a wide range of journals. These can be found by using the various indexing services, including *Geographical Abstracts*, *Public Affairs Information Service*, *Social Sciences Index*, or the *British Humanities Index*.

FURTHER READING

1949 to 1976

There is a voluminous literature on this period. A good sense of the contrasting viewpoints and interpretations can be got from the following studies:

Howe, C. (1978) *China's Economy: A Basic Guide*, London: Paul Elek.
    A careful, analytic, factual survey of the Chinese economy.
Maxwell, N. (ed.) (1979) *China's Road to Development* (second edition), Oxford: Pergamon.
    Essays on Chinese development largely from positions sympathetic to Maoism. Contains an excellent biblography on Chinese development up to about 1976.

1976 to 1989

There is a growing literature on the policies of the post-Maoist governments. These include:

Benewick, B. and Wingrove, P. (eds) (1988) *Reforming the Revolution*, London: Macmillan.
    A basic guide to the reform decade aimed at students and lay readers, with each chapter summarising the new policies and their impact.
Burns, J. and Rosen, S. (1986) *Policy Conflicts in Post-Maoist China*, Armonk, New York: M.E. Sharpe.
    Translations of Chinese documents illustrating conflicts over ideology, industrial and agricultural policy, and population.
The China Quarterly: *The Readjustment in the Chinese Economy* (1984), London: Contemporary China Institute.
    Analytic essays on macro-economic policy, population, industry, and agriculture. This is a special theme issue of the journal for December 1984, released as a book.
Chossudovsky, M. (1986) *Towards Capitalist Restoration? Chinese Socialism After Mao*, London: Macmillan.
    A powerful critique from a 'Maoist position', highly critical of Deng Xiaoping's opening of China to foreign capital.
Gittings, J. (1989) *China Changes Face: Socialism and Reform 1949–1989*, Hong Kong and Oxford: Oxford University Press.
    A book which is both readable and extremely well-informed, and which helps understand the shifts in policies over the 40-year period.
Goodman, D.S.G. (1989) *China's Regional Development*, London: Routledge.
    A collection of essays on the political and financial issues underlying regional differences and conflicts, together with surveys of the problem sectors of energy and transport, and case studies of different regions.
Joint Economic Committee, Congress of the United States (1982) *China Under the Four Modernisations* Parts I and II, Washington: U.S. Government Printing Office.
    Essays on many aspects of Chinese economy, society and politics since 1976.
Reynolds, B.L. (1987) *Reform in China: Challenges and Choices*, Armonk, New York: M.E. Sharpe.
    Reports by Chinese scholars of the extent of reforms (as of 1985) and what directions policy should take.
World Bank (1983) *China: Socialist Economic Development* (three volumes), Washington: World Bank.

Analysis of various aspects of Chinese economy, based on information made available to the Bank by the Chinese government.

World Bank (1985) *China: Long-Term Development Issues and Options*, Baltimore: John Hopkins University Press.
More 'popular' and reflective analysis than above. Both represent an important index of China's increased involvement in the 'capitalist world economy'.

Atlases and other books of interest to geographers

Blunden, C. and Elvin, M. (1983) *Cultural Atlas of China*, Oxford: Phaidon.
Superbly produced introduction to Chinese history and culture illustrated with maps.

Buchanan, K. (1970) *The Transformation of the Chinese Earth*, London: Bell.
Very dated but worth reading critically for its sympathy for Maoist China, and a general geographical introduction to the country.

Cole, J. P. (1985) *China 1950–2000; Performance and Prospects*, Nottingham: Department of Geography.
A cyclostyled survey, particularly valuable for its statistical analysis, innovative cartography and value judgements on China's future.

Dwyer, D. (ed.) (1974) *China Now*, London: Longman.
Very dated, but particularly useful for its reprints of 'classic' Western analyses of China's geography.

Money, D. (1990) *China: The Land and the People* (second edition), London: Evans.
A textbook aimed at upper secondary students but deserving more widespread use for its clear text and excellent maps and colour photographs.

Pannell, C. and Ma, L. (1983) *China: The Geography of Development and Modernisation*, London: Edward Arnold.
A systematic textbook aimed at degree-level students.

The Times (1974) *The Times Atlas of China*, London: Times Newspapers.
Includes historical and physical maps, maps of provinces and major cities.

Tregear, T. R. (1980) *China: A Geographical Survey*, London: Hodder & Stoughton.
Beware the 1980 publication date: this is essentially a recasting of books published in 1965 and 1970.

Wood, F., Siven, *et al.* (1988) *The Contemporary Atlas of China*, London: Weidenfeld & Nicholson.
Area maps plus short essays on the regions of China, life in town and country, religion. More an encyclopaedia than an atlas, but very useful for this.

CHINESE ECONOMIC AND POLITICAL CHANGES 1949–89

1 October 1949 The People's Republic of China is declared by Mao Zedong after two decades of civil war and resistance to Japanese invasion.

1949–52 Land reform Aimed at freeing the agricultural base of the economy from the exploitative system of landlord and rich peasant domination. Land confiscated from landlords and rich peasants and distributed to poorer classes of peasants.

1952–8 Collectivisation Begun by promoting mutual aid teams, comprising several households which worked together to compensate for shortages of draught animals and farm implements. Followed by various stages of collectivisation using co-operatives.

1953–7 First Five-Year Plan To develop a widespread industrial structure. Priority given to producer goods – i.e. steel, coal, cement – in order to facilitate industrial expansion. Considerable Soviet technical assistance received. Little state investment went into agriculture, which was being grouped into co-operatives. Agriculture did not develop fast enough to raise rural living standards or provide sufficient surpluses of food and raw materials for the urban population and industry. Slow growth of light industry meant fewer consumer goods to provide incentives for agricultural producers.

1958–60 The Great Leap Forward (GLF) A crash industrialisation programme throughout the economy. Based on the 'walking on two legs' policy for agriculture and industry, it involved their simultaneous development, promoting small- and medium- as well as large-scale industry. It coincided with a movement to set up *People's Communes* to organise agricultural production and administer rural China. The communes were formed from existing co-operatives to make up larger-scale units. Divisions with the USSR, and Soviet advisors leave and 'tear up contracts' for the heavy industry projects in which they were engaged.

1959–61 Famine disaster Economic chaos caused by GLF compounded by flood and drought lead to 13 and 26 million famine deaths. Sharp political and ideological struggles over policy directions in agriculture and industry result.

1961–4 Readjustment Greater scope for market mechanism to influence peasant production (including private plots) and improve the economy. Liu Shaoqi prominent in economic policy.

1966–76 The Cultural Revolution Emphasis on ideological struggle rather than economic incentives as a basis for economic progress. Mao Zedong's faction of leadership encourages mass movement to overthrow so-called followers of the 'capitalist road' from positions of authority. During 1969–76, despite slow reassertion of influence by the pragmatists (after 1969), the struggle continues in its various spheres.

8 January 1976 Premier Zhou Enlai dies after a prolonged struggle against cancer.

28 July 1976 A massive earthquake centred on Tangshan kills a quarter of a million people.

9 September 1976 Mao Zedong, Chairman of the Communist Party, dies after prolonged illness.

6 October 1976 Hua Guofeng orders the arrest of the 'Gang of Four', a group of 'ultra-left' radicals with a political base in Shanghai, and including Mao's wife, Jiang Qing.

1976–8 Socialist modernisation Mao Zedong's death and the arrest of the Gang of Four permits the emergence of a new interim leadership, which is rather unstable. New emphasis on Four Modernisations in agriculture, industry, science and technology, and defence, to raise the standard of living. Increasing role of Deng Xiaoping.

February 1978 The Fifth National People's Congress meets, and it begins to regain some of its constitutional powers as China's supposed governing body.

December 1978 After crucial struggles over the culpability of Mao Zedong in relation to the Cultural Revolution, the Communist Party holds the Third Plenum of its Eleventh Central Committee, making economic growth the priority. Deng Xiaoping becomes leading individual promoting the reform policies.

1979–80 Reforms begin Change from correct politics as the measure of success, to stress on economic progress. Priorities in the economy altered; emphasis is now on light industry to provide consumer goods. Greater availability of these is needed if the 'new responsibility' principle, which provides wage incentives to encourage higher productivity, is to work. Market principle is given greater sway and profits seen as a valid tool for assessing viability and progress in enterprises. The 'single child' policy is introduced.

1979 The year witnesses a wide range of policy reversals, in particular the introduction of the rural responsibility system. Joint ventures with foreign firms are permitted and the four Special Economic Zones established. In population control, the policy of limiting parents to a single child is introduced. The flowering of much more open debate on political and cultural issues in the 'Peking Spring' of early 1979, symbolised by the so-called 'Democracy Wall' in various cities, is soon ended by arrests and repression.

1980 State-run industrial sector begins experiments which lead to enterprises retaining a larger share of profits, the introduction of bonuses for workers, and more independent decision-making. Important changes are made in the political leadership, and Hua Guofeng 'resigns' as premier to be replaced by Zhao Ziyang. Another of Deng's protégés, Hu Yaobang, is appointed party General Secretary.

November 1980 The trial of the 'Gang' begins; with them are also charged other 'ultra-leftists' including Mao's former secretary, Chen Boda.

1981–5 Sixth five-year plan Priority given to consumption over investment, light over heavy industry, and agriculture over industry. Greater integration into international trade system, and less emphasis on self-reliance, as foreign technology is increasingly seen as the basis for modernisation. Communes abolished, family farming (using the responsibility system) introduced.

July 1982 The first census to be conducted since 1964 shows China to have one billion people.

1984 In addition to the SEZs (announced in 1979), 14 coastal ports are designated as 'open cities' (with similar incentives to capital investment from abroad) with moves to improve infrastructure for better import and export trade.

October 1986 The Queen visits China, including Hong Kong, the first British monarch to do so.

1986–90 Seventh five-year plan Greater priority to quality, technological transformation and balancing supply and demand. Increased priority to energy and transport sectors. Trade to be expanded and priority to established coastal areas that can benefit from export-orientated strategy.

1989 Democracy demonstrations, martial law and massacres Uncertainty as to what will happen to the economic reform programme. The immediate post-massacres leadership claims to be committed to continuing the reforms, but is known to include critics of the impact of the market philosophy and some who espouse more central direction of the economy.

Chapter two

Regions: spatial inequality and regional policy

Terry Cannon

The social and economic characteristics of China vary tremendously from place to place, and the contrasts are so sharp that they surprise people who do not know the country well. For instance, even the average per capita consumption of Beijing city-dwellers is nearly five times higher than that of the peasants of Gansu, one of the poorest provinces.

This chapter looks at some of the patterns inherited by the new communist government in 1949, and how it has influenced the spatial patterning of economic activity and people's income and welfare since 1949. It examines how far since 1949 (and especially since 1979) the Communist Party's policies have responded to perceived inequalities, and the degree to which policies have had a deliberate or accidental geographical impact.

It is argued that only occasionally has the leadership shown concern to reduce certain forms of spatial inequality: the Communist Party of China (CPC) has been motivated much more by military strategy or its notion of economic efficiency. It has therefore been more interested in determining where economic activity should be located, rather than addressing income or welfare differences.

This should not be surprising: the revolution was not aimed at removing all inequalities, but rather to end class exploitation and foreign domination. Many geographical differences are not the result of exploitation, arising from natural conditions (like climate), or from living in locations which enjoy an advantage (for instance, in access to markets for the sale of produce).

These have not often been a communist priority, since it can easily be assumed that they are not the result of one class enriching itself at the expense of another. Dealing with such differences has to await the greater affluence of fully developed communism, at which time society can share on the basis of need. In the meantime, limited programmes of aid to poorer districts have been used, as well as disaster aid after the effect of hazards.

INEQUALITIES IN SPATIAL AND REGIONAL TERMS

Until the 1980s, the Chinese state had a virtual monopoly in providing information about the country to the world, and its quality was not good. The predominant image of the country was one in which party policy could be implemented throughout the land, and of a society virtually homogeneous in its economic and social characteristics as a result of this all-pervading presence. Independent travel was difficult, so foreigners went mainly in organised groups to a few places, and gained an impression (or wanted to believe) that things were much the same everywhere.

Since 1979 much more (and better) information has become available, and from more sources than the government alone. There is now greater awareness of spatial differences within the country, even if these are mostly seen in terms of very general contrasts between rather ill-defined regions (especially between 'the Coast' and the rest). Even so, most data was available only at the level of China's twenty-nine provinces, and this made careful regional analysis impossible.

Fortunately, in 1982 there was a detailed census from which there are statistics relating to China's more than 2,000 counties. These allow a much finer understanding of spatial variations, which are shown to be much more complex than can be represented on a 'regional' basis using simply provinces. Figure 2.1 is a simplified map showing the per capita gross value of industrial and agricultural output on a county basis. This data is of production for the two main sectors of the economy, not of income, but is a useful measure of relative levels of wealth across the country. One thing which is apparent is the high level of variation, and the lack of any clear, simple regional pattern.

But such data provides only part of the picture: regions are not the only form in which spatial differences show up. Of crucial 'non-regional' importance, for instance, is the dynamic which connects and separates town and countryside. This is dealt with rather more in the chapters on rural China (Chapter six) and urbanisation (Chapter eight), while here the main focus is on larger-scale regions.

This chapter explores the economic, political and social processes in China and the world which have had a spatial impact in China. Unfortunately, despite the data available from the census, it is done in highly simplified macro-regional terms. In fact, for much of the discussion, reference is made to just three very large regions: the Coastal, Central and Western (Figure 2.2). This is far from adequate, given their high degree of internal differentiation, but it is necessary here for simplification, and more especially because it is by reference to these three that the Chinese themselves now debate regional issues and define regional objectives. Before looking at them in more detail, there are a number of other spatial matters to mention.

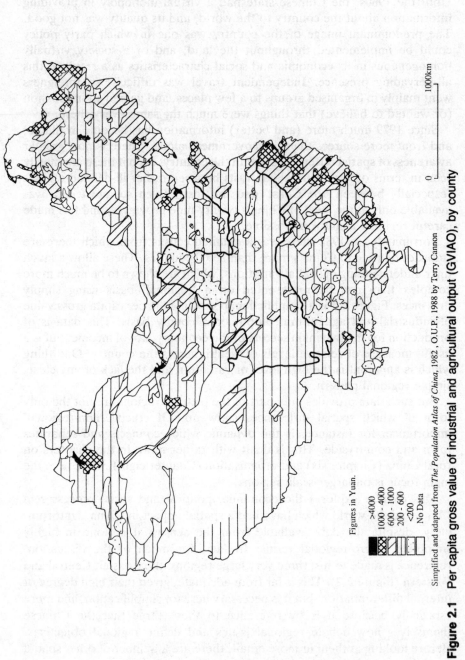

Figures in Yuan.

>4000
1000 - 4000
600 - 1000
200 - 600
<200
No Data

0   1000km

Simplified and adapted from *The Population Atlas of China*, 1982, O.U.P., 1988 by Terry Cannon

Figure 2.1 Per capita gross value of industrial and agricultural output (GVIAO), by county

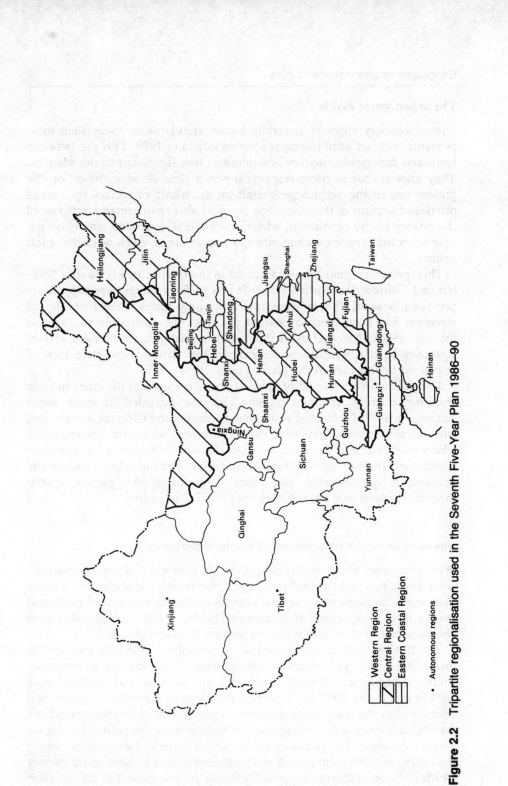

Figure 2.2 Tripartite regionalisation used in the Seventh Five-Year Plan 1986–90

Western Region

Central Region

Eastern Coastal Region

• Autonomous regions

Heilongjiang

Jilin

Liaoning

Jiangsu

Shanghai

Zhejiang

Taiwan

Beijing

Tianjin

Shandong

Anhui

Jiangxi

Fujian

Inner Mongolia

Shanxi

Hebei

Henan

Hubei

Hunan

Guangdong

Ningxia

Shaanxi

Gansu

Sichuan

Guizhou

Guangxi

Hainan

Qinghai

Yunnan

Xinjiang

Tibet

The urban–rural divide

Urban dwellers enjoyed generally higher standards of living than most peasants, at least until the rural reforms begun in 1979. This gap between town and contryside, worker and peasant, was significant to the Maoists. They argued that in some respects it was a class division, based on the greater use of the surplus generated by the whole of society by a small privileged section in the cities. For them, it also represented a reversal of the history of the revolution, whose success had depended mainly on the help from the farming communities in poorer areas which were being left behind.

This spatial inequality was addressed in the Great Leap Forward (1958–60) and Cultural Revolution (1966–76) by policies which were supposedly pro poor peasant. These were really the only times when such special attention has been paid to peasant priorities for egalitarian reasons. In the mid-1960s, Mao rebuked the health ministry for running a system favouring 'urban gentlemen' because of the virtually complete lack of organised health care in the countryside.

The policy he developed of medical teams going from the cities to train 'barefoot doctors' (peasant paramedics) was extended to much wider 'sending-down' of educated young people during the Cultural Revolution, with the intention of increasing rural education. As is now known, many who went then, and the older intellectuals who followed, were unwilling participants in this Cultural Revolution policy, and the value it had for the peasants is questionable. But other than during that period, spatial inequalities have not figured very high in CPC priorities.

The state as major determinant of spatial differences

What is unusual about spatial inequalities in China is that the government, since 1949, has had the ability to shape the country spatially much more than most. Because of its central control of the economy, and command over a large proportion of investment funds, it has the capability (not always used) to alter differences in welfare between places.

Yet this control is not universal or complete: there is also spatial inequality in the government's effectiveness. The country is immense, communications are difficult, and there are political and administrative weaknesses in the CPC itself. During the Cultural Revolution, nearly two decades after the party came to power, a group of Red Guards arrived in a remote valley in the southwest, the first visitors from the political realm for a very long time. The peasants who received them had a question: who is the Emperor? The local people were aware that they lived in territory over which Chinese emperors supposedly ruled, yet the impact of this on their lives for decades had been practically non-existent.

This is an extreme case, but illustrates quite how diverse the country is. Spatial variation is influenced by many factors, including natural conditions of climate and resource endowments, and the various sorts of economic system which people have come to operate in different places. These vary enormously, including cultivation in oases in the far west, slash-and-burn farming in the southwest, as well as the more familiar intensive farming and industrialism of the east. To these factors must be added the actions (or inaction) of government.

The inadequacy of simplified regions

As a result of all these, China is an enormously intricate patchwork of social and economic variation, not easily subsumed into a few large regions. The use here of the three simplified regions designated by China is inadequate, and needs to be qualified so as not to assume too much similarity within them.

For instance, income differences can vary greatly over quite short distances (even between neighbouring villages). It is evident (from 1982 census data – see Figure 2.1) that some counties less than 250 km from Shanghai, in what is often considered the 'wealthy' Coastal Region, are amongst the poorest in the country. Conversely, some of those which have highest per capita agricultural incomes are in Tibet and Xinjiang, which are part of the 'poor' Western Region (see Figure 2.2).

The ways that regions are conceived can vary enormously. An area of territory may be considered a region by virtue of its apparent internal sharing of chosen characteristics (e.g. topographic or socio-economic) which make it distinct from those of neighbouring territory. Alternatively, a region may be designated by the state for the purpose of generating certain shared characteristics or other policy objectives within a given area of territory, which if achieved would distinguish it from the surrounding areas.

The three macro-regions represent a type of region which both identifies certain crude, shared, internal characteristics and then links them to the setting of broad objectives in economic development. These were formally set out in the Seventh Five-Year Plan for 1986–90, and are discussed in that context later. Meanwhile, they will be used to provide a much simplified understanding of change in China's economic development since 1949.

REGIONS IN THE REVOLUTION

When the communist government came to power in 1949 the country had been at war within itself or against the Japanese for more than thirty years. The legacy of tremendous destruction and disruption of the economy was overcome remarkably quickly, especially given the lack of experience of the CPC in managing large industry and cities.

Inherited inequalities

Virtually all industrial activity in 1949 was located in two types of area, both now considered to be part of the Coastal Region. These were, first, the disconnected and highly localized developments in cities – all of them ex-treaty ports – on the coastal fringe (e.g. Shanghai, Tianjin, and Qingdao), where foreign entrepreneurs and their Chinese emulators had established various businesses after 1840 (see Chapter three). Second, in the northeast provinces which collectively became known as Manchuria, Russian and later Japanese colonisation led to significant development of raw material resources, especially iron and coal, and heavy industry, much of it serving the needs of the Japanese economy.

The rest of the country was virtually devoid of modern industry: Beijing (Peking), though not very distant from either of the areas mentioned, had virtually none. Other cities which today are notorious for their industrial pollution, many of them far inland like Taiyuan, Xian, Wuhan, Chengdu, Lanzhou, or Kunming, were innocent of such disturbances. This does not mean that China had never had industry. In many respects the nation had been more advanced than anywhere else in the world in Song times (in the twelfth and thirteenth centuries AD), and some industries (employing many thousands) were continuously successful long after that. But agriculture continued to dominate, and even the inroads of foreign-initiated industry after 1840 made little impact. China remained having but 'a modern hem on an ancient garment'.

The First Five-Year Plan and its aftermath

The system of central planning begun in China in the First FYP of 1953–7 owed much to the Soviet experience, and was indeed largely financed and directed by loans and technicians from the USSR. This 'command economy' established a large number of important enterprises in sectors such as iron and steel, metallurgy, chemicals, electricity generation, coal mining and textiles. In short, the emphasis was on large-scale plants, and a lot of them in heavy industry and under strict central control.

The plan partly reflected the Soviet Union's domination over communist politics. But it was also a product of the international environment with which the new Chinese state found itself having to cope (see Chapter eleven). In this, strategic concerns determined both the nature of rapid accumulation, and the location of investment.

The very nature of the industry on which the plan concentrated also had its spatial effect. Much of it relied on coal and other locationally specific

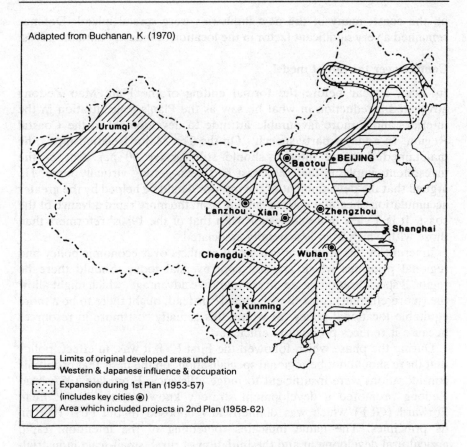

Adapted from Buchanan, K. (1970)

Limits of original developed areas under
Western & Japanese influence & occupation

Expansion during 1st Plan (1953-57)
(includes key cities ⊚)

Area which included projects in 2nd Plan (1958-62)

Figure 2.3 The inland spread of industry in the First and Second Five-Year Plan

raw materials, and so was concentrated into relatively few (but large) centres (known as the Key Point cities). This meant that the benefits of the new jobs were spread beyond the Coastal Region, but to relatively few places in the other regions. Being spread so thinly, this industrial growth hardly constitutes the creation of a new regional pattern. But the impact was to disperse a significant amount of industrial capacity (Figure 2.3). Of nearly 700 large- and medium-scale projects underway in the First FYP, two-thirds were inland.

In the post-1949 'Cold War' atmosphere, China was effectively embargoed by the Western powers, and so self-reliance and acceptance of Soviet aid was a result of constrained options rather than ideology alone. The Korean War (1950–3) and its aftermath put China on a war footing, and this also affected the type of industries which were given priority. But this also determined that military-strategic concerns affected the location of the new projects. To reduce risks from possible bombing and invasion on the coast, many of the new industries were spread inland. Defence remained a very significant factor in the location of industry until the 1970s.

Conflict over the Soviet model

In 1956, a year before the formal ending of the Plan, Mao Zedong advocated a reduction in what he saw as the Plan's concentration in the interior, and a more favourable attitude to development in the Coastal Region. In a major party speech ('On the Ten Major Relationships'), he maintained that the interior should still receive 90 per cent of the investment, implying that the coast had been getting virtually none. He argued that the development of the interior would be helped by the greater accumulation of capital made available from the more rapid advance of the coast. It is a view more consistent with that of the 1980s' reformers than those with which Mao is generally associated.

Inherent in this argument is a set of conflicts over economic policy and regional priorities which are still controversial today. Should there be regional specialisation based on comparative advantage, which might allow the (more efficient) coast to race ahead? Instead, ought there to be a more equitable locational policy, or would this actually cost more in resources because it reduces production efficiency?

During the phase which followed the First FYP it was, in effect, policy that there should not be regional specialisation, on the basis that economic considerations were insufficient to judge efficiency. From late 1957, Mao Zedong promoted a development strategy known as the Great Leap Forward (GLF) which was diametrically opposed to the First FYP in its principles. The name indicates something of the intention: rapid agricultural development and the initiation of rural, small-scale industrialisation, with all areas attempting local self-sufficiency in some key industrial products.

In theory the GLF (and the concurrent commune system) was intended to promote a greater spread of economic opportunity throughout rural China. This was to be, even at the expense of 'efficiency' as normally defined. But to understand the GLF once again requires an appreciation of the international situation, and the growing threat after 1958 in the south of China. Taiwan was increasingly belligerent towards the People's Republic at the same time as the USA was planning to base nuclear missiles on that island. As if this were not enough, antagonism with the USSR led in 1960 to the Soviet Union withdrawing its aid, and the beginning of the thirty-year rift.

The 'Third Front' and strategic factors in regional development

Even after coming to power, and with national defence much in mind, the CPC government retained a form of strategic regionalisation which divided the country into six military regions, each with its headquarters in a major provincial city. It was based on the idea of regional self-sufficiency in the production of key minerals and industrial products, so that in event of civil war or invasion there were smaller units which supposedly could survive independently. The nature of external threats changed in the 1960s and required another approach. The military threat was from both super-powers, so where could economic and military activity now be placed for security?

A quite extraordinary episode in regional development now emerged, lasting from about 1965 to the early 1970s. We now know that over a seven-year period there were massive and secret investments in certain parts of the country which were thought safest from invasion and bombing, known as the Third Front (*sanxian*) region in the southwest (Figure 2.4). Such was the nature and size of these developments and their associated transfers of personnel that the effects on the efficiency of the economy are likely to continue to be felt for some time to come.

The answer to the threat of attack from either or both the USA and the USSR was a policy to duplicate or withdraw many industrial enterprises, research establishments and military installations into this interior heart-land. The first front was regarded as the highly vulnerable coastal cities; the second a rather vague 'intermediate zone' between the first and third.

The core province of the Third Front was Sichuan, an inland basin of rich agricultural land, relatively well off in fuels, and surrounded by mountains. The neighbouring provinces or parts of them were also involved (Guizhou, Yunnan, Shaanxi, Gansu, and the western parts of Henan, Hubei and Hunan). Further west in Tibet, Qinghai and Xinjiang, other military and nuclear establishments were built up or expanded under this policy, making eleven provinces involved in all.

The *sanxian* policy relied on safeguarding crucial productive capacity and research facilities in order to prosecute a war from a protected heartland. Its location was dictated by the threat from all sides, and the resultant industrial development is significantly different from the First FYP strategic policy, which had no need to avoid places near the Soviet and Mongolian borders.

Although US President Nixon's diplomatic openings with China in 1972 reduced the threat of US aggression and complete hostile encirclement was broken, very little has been known about *sanxian* outside China until relatively recently. The story emerged because of the current need to use the technology and personnel locked into the secret economy for civil purposes. Today the international situation hardly warrants such a strategic measure in any case, and China is more interested in using the productive and research capabilities of the Third Front establishments for modernising and developing the rest of the economy.

The revelations about the policy show an astounding scale of investment,

USSR
PEOPLES REPUBLIC
OF MONGOLIA
USSR
INNER
MONGOLIA
A.R.
HEILONGJIANG
JILIN
LIAONING
Urumqi
XINJIANG UYGUR
A.R.
Hohhot
HEBEI
Beijing
GANSU
SHAANXI
SHANXI
Taiyuan
Xining
QINGHAI
Lanzhou
SHANDONG
Zhengzhou
JIANGSU
TIBET A.R.
Xian
HENAN
HUBEI
ANHUI
Lhasa
Chengdu
SICHUAN
Changsha
Guiyang
HUNAN
JIANGXI
FUJIAN
GUIZHOU
Kunming
GUANGXI
YUNNAN
Nanning
GUANGDONG
TAIWAN
0
800 kms
Uncertain extent, indicates other important installations; ie : nuclear
Solid line encloses known Sanxian area
Shaded area indicates the Sichuan core, receiving one-third of San Xian investment
A.R.
Autonomous Region
The approximate limits of the San Xian strategic region of the 1960's and early 1970's, where many military and key industrial centres were located away from the Soviet Borders and the coast.

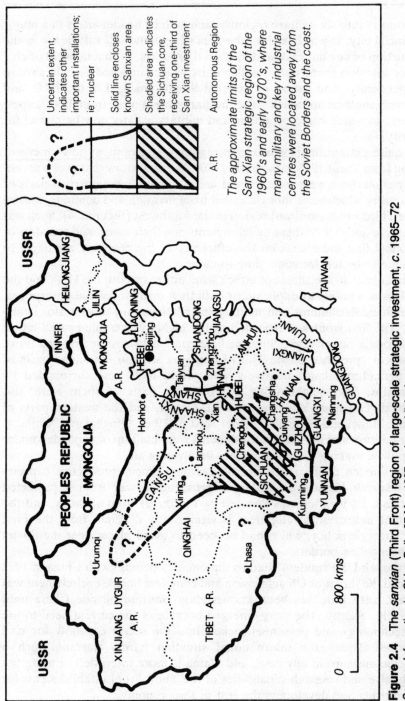

Figure 2.4 The *sanxian* (Third Front) region of large-scale strategic investment, *c.* 1965–72
Source: Information in *China Daily*, 27 May 1987 and unpublished papers.

involving billions of yuan on several huge key projects. That China's fifth largest iron and steel works today is at Panzhihua, on the borders of Sichuan and Yunnan in a very mountainous area, gives some idea of the scale and nature of the policy. The *Guardian*'s China correspondent reported in 1988 that Panzhihua, which began from scratch twenty years ago, with the first workers living in huts without sufficient food, is now a city of more than 850,000 people.

Most staggering of all, though, are the amounts of capital invested in the Third Front region. About 29,000 institutions of all sizes were involved, and of these around 2,000 were large and medium-sized enterprises (in all China there were only 5,000 so defined in 1981). If we look at the eight provinces nearer the heart of the region (i.e. excluding Tibet, Qinghai and Xinjiang), their share of total national investment is put at nearly 53 per cent in 1966–70, and 41 per cent in 1971–5. By contrast, their share of the national total investment made in 1985 in fixed assets was only 24.6 per cent.

One study of this strategic region asserts with reason: 'We can safely conclude that at least two-thirds of budgetary industrial investment went to the Third Front during its prime construction period' (Naughton 1988: 351–86). The population in this region is around 38 per cent of China's total, suggesting that the massive 'excess' investment (when compared with the share of population) in the *sanxian* period has in the 1980s been reversed to considerable 'under-investment'.

The dispersal of enterprises and institutions under the *sanxian* policy had little at all to do with bringing about spatial equality. Interpretations which assume that Chinese policy has been concerned with this are misguided. At times in CPC debates the question of greater regional equality has been raised, but the evidence is that major investment shifts, which have indeed had such massive regional effects, have been almost entirely for strategic reasons. The CPC has undoubtedly got the capacity to bring about major shifts in resources between regions, and on a number of occasions has done so. But the motivation has been to strengthen and preserve the country, not to equalise consumption or welfare between provinces. All analyses of the country's economy until very recently have had no knowledge of this factor and so are fundamentally flawed.

POST-MAO ECONOMIC REFORM AND ITS REGIONAL IMPACT

The post-Mao era since 1979 has been marked by a major shift in emphasis, away from 'politics' to the promotion of 'economic' priorities, embodied in the Four Modernisations (of industry, agriculture, science/technology and defence). This does not mean that China is devoid of politics. But political struggle and the time-consuming discussion of what is politically correct should no longer handicap the production process (see

Chapter one). Each of the modernisations has been promoted by a series of economic reforms, of which those affecting rural production were among the earliest (begun in 1979).

Although many rural people enjoy the higher living standards which have accompanied the reforms (see Chapter six), there is significant spatial inequality in the reforms' impact. Significant poverty and hunger exists in areas officially acknowledged to include 100 million people. This requires the state to provide limited aid to certain counties.

Urban reforms for industrial growth

'Urban' reforms, directed at reorganising industry, were more difficult to implement. Their key characteristic is reduced central control over the economy. At lower levels of the administrative hierarchy, where previously orders were received from superior authorities, there is now much greater autonomy and new financial arrangements which reduce dependence on the centre. But there has been significant opposition from old-guard CPC members who still favour centralised control.

This decentralisation has had regional and spatial objectives associated with it, for instance the encouragement of regional comparative advantage as opposed to regional self-sufficiency. In parallel with the reform of industry have come policies to encourage new spatial forms in the economy, such as the designation of regional associations of enterprises which cut across province boundaries. Instead of the plan emphasis on hierarchy and vertical integration of enterprises, the reforms encourage 'horizontal linkages'. Additionally, the post-Mao policies include the 'open door' to foreign investment and trade, and this has also had its regional impact in its effects on industry and commerce.

So, in various ways, there are new regional strengths and weaknesses emerging, and these are contributing to the economic and political crises which have affected the success and continuity of the reform period. Most serious of these, as we shall see, are the growing protests at the 'gap' between the west and the coast, and the way in which local (especially provincial) economic power has disrupted the objectives of reform and led to serious conflicts between areas.

Regional comparative advantage

China's leaders have argued in recent years that regional or local self-sufficiency is inefficient and that instead the country must pursue a policy of allowing some areas to 'get rich first'. As a 1985 article in *Beijing Review* put it: 'China is vast, and its natural and production conditions vary from place to place. ... Therefore the pace at which areas and peoples become prosperous will never be simultaneous.' This is a clear spatial implication of the economic 'reforms' of the past decade, and is related to the increased role granted to the market. It is analogous with the policy that some individual peasants or entrepreneurs should be allowed to become rich before others.

It is really a policy of regional comparative advantage: particular areas are thought more efficient than others at producing some things, so they should specialise by exploiting their cost advantage. The surpluses and deficits which arise for the different products of each should be equalised through increased inter-regional trade (despite the inadequacies of the transport network). This contradicts the 1950s' and 1960s' policies, which were aimed at regional self-sufficiency in order to reduce transport needs and improve defence capabilities.

Such regional specialisation is meant to be a result of the operation of market forces, at least in part. But the government also has a clear notion of what pattern it prefers, based on what it sees as existing material conditions. This preference is demonstrated in the way the three macro-regions have been drawn up in the Seventh FYP (see Table 2.1 and Figure 2.2). Each of the three is described in terms of its existing resources, and the economic priorities it is supposed to pursue in the next ten years or so. It is the explicit promotion of the rapid advance of the Coastal Region in this which has resulted in conflict and protests from the leaders of both Central and Western provinces. Although some basic characteristics and priorities for the three regions are given in Tables 2.1, 2.2 and 2.3, it needs to be remembered that there is much internal variation within each of them.

Decentralisation and the dilution of planning

The central planning system was already largely decentralised during the 1970s in the sense that the Beijing authorities did not try to run everything.

Table 2.1 Macro-regions of the Seventh Five-Year Plan: proportion of population by region compared with share of investment, 1985

Region	Population		Total investment in fixed assets[a]	
	Millions	Per cent	RMB[b] millions	Per cent
Coastal	429.96	41.3	127,483.0	52.7
Central	371.25	35.7	74,675.0	30.9
Western	239.87	23.0	39,597.0	16.4
Total	1,041.08		241,755.0	

Source: Statistical Year Book of China 1986, (1987) Hong Kong: Oxford University Press, p. 73 (population), p. 367 (investment).
Notes: [a] 'Total investment in fixed assets' is a category which shows the amount of work done in construction and purchase of fixed assets by state-owned, urban and collective units and individuals, expressed in money terms.
[b] RMB = Renminbi, the official name of the currency unit yuan (¥).

Table 2.2 Macro-regions of the Seventh Five-Year Plan: proposals for specialisation and differential development of the regions

COASTAL

- Technological updating of existing industries and creation of new industries, especially in high value and consumer goods, using central-government investment and foreign partners.
- Special Economic Zones, open coastal cities and other areas to grow rapidly and become bases of expanded foreign trade, and act as training grounds for transmitting knowledge and technology to other parts of the country. Such areas to continue preferential policies.
- Energy resources and transport to be developed to ease shortages. Heavy users of energy and transport to be closed or relocated in Central Region.
- Agriculture and rural production to be expanded and linked more to urban and industrial demands.
- Service sector to be expanded, including financial services and the setting-up of markets and trading centres to encourage circulation of manufacturers and farm produce, and rural inputs, imports and exports.
- Preferencial treatment for export industries, and promotion of increased tourism.

CENTRAL

- Speed-up of coal and electricity generating, non-ferrous metals and phosphate mining, and building-materials industries, using increased government investment.
- Cities and areas which are more developed will have greater development of hi-tech industry, both existing and new.
- Vigorous development of agriculture ro raise output of grain and cash crops.
- Accelerated development in coal and electricity and oil, and increased capacity for steel, using government and foreign investment. Also more transport links to coast.
- More rapid introduction of new technology in existing industry.
- Setting-up of 'commodity production bases' for grain, soy bean, oil-bearing seed crops and sugar-yielding crops.
- Vigorous development of forestry, animal husbandry and animal prodcts.
- Development of the middle reaches of the Changjiang (Yangzi) river to help stimulate growth in the Western Region. (It is not specific whether this means the Three Gorges hydro-electric project.)

WESTERN

- Development of farming, forestry, animal husbandry and transport.
- Increased exploitation of minerals and energy supplies on a selective basis.
- Where economic and technical level is relatively high, technological up-dating to be undertaken, especially using co-operation with Coastal and Central Regions.
- Land for grain production to be protected, to raise yields and reduce imports from other regions.
- Faster development of grasslands and pastoralism, but with environmental protection.
- Improved rail links with the Central and Coastal Regions.
- Transfer of defence industries to civilian production.
- Open up mineral and energy sources in the higher reaches of both Huanghe (Yellow) and Changjiang (Yangzi) rivers.
- Development of the Sichuan–Yunnan–Guizhou border area in energy and semi-finished products.
- Develop the Urumqi–Karamai area as an industrial centre in Xinjiang.
- Faster construction of frontier market towns, and expanded foreign trade across borders in west.
- Preferential treatment for the west in development of education, transport, mining and energy.

Source: adapted from 'The Seventh Five-Year Plan of the People's Republic of China for Economic and Social Development (1986–1990)', *Beijing Review* 29, 17: 28 April 1986.

Table 2.3 Comparison of the value of production in industry and agriculture in the three macro-regions in 1985

Region	1985 (billions of yuan)	Increase over 1980 (per cent)	Percentage of total production 1980	1985
Value of industrial output				
Coastal	56.64	79.5	60.0	61.2
Central	24.57	71.6	27.3	26.5
Western	11.34	70.5	12.7	12.3
Total	92.55	76.2	100.0	100.0
Value of agricultural output				
Coastal	12.62	51.8	42.8	43.3
Central	10.63	49.5	36.6	36.5
Western	5.87	46.3	20.6	20.2
Total	29.12	49.8	100.0	100.0
Value of industrial and agricultural production				
Coastal	69.27	73.7	55.4	56.9
Central	35.2	64.3	29.8	28.9
Western	17.21	61.4	14.8	14.2
Total	121.68	69.1	100.0	100.0

Source: 'Economic growth in different areas', Beijing Review 29, 49: 8 December 1986.

State-owned industry was delegated to the provinces and even to some city governments. But the crucial issue was who controlled profits and investment funds. Before 1979 these stayed firmly in the hands of the centre; since then, these too have been decentralised, and the centre allows local-level bodies to hold on to more of their profits and pay a tax to the centre instead.

It is this transfer of control over funds which has largely been responsible for the boom in the industrial sector since 1980, with growth rates of well over 10 per cent per annum, much higher in some years. The authorities which have gained more control over profits and capital are not so much the provinces but local authorities, including cities, townships and counties. These lower-level governments have, in effect, gone into business, often in a big way. Enterprise culture in China is based on bureaucrats and local party chiefs turned capitalists, and much of this growth is in rural areas.

The post-Mao government has also reduced the rigid 'central' planning, which it regarded as a constraint on the economy, by the devolution of power to enterprises themselves; though combined with the exercise of much greater control in their sphere by local levels of government, a tension has been created between the objectives of enterprise managers and the local bodies which oversee them.

In effect, the urban reforms mean the loss of control of much of the economy by central government. This is evident in a number of ways beyond the intended decentralisation, including the deliberate flouting by lower-level authorities of government directives and regulations. Some of these had been intended to minimise the damaging effects of the process. Let us look at some of the implications of decentralisation at different levels of the hierarchy.

Enterprises

With more autonomy granted to individual state-owned enterprises under the urban reforms, which were mainly introduced after 1984, industry has become more 'footloose'. And whereas before the emphasis was on manufacturing, other activities in commerce and the tertiary sector have been experiencing massive growth. Enterprises and local authorities now retain more investment funds, and are thus able to reinvest in other locations, and in sectors unrelated to their previous activities. Before, most funds were channelled through central authorities and had to be used within the enterprise (see Chapter seven).

This has led to cases of investment in places far away from the originating enterprise (in a few instances even abroad). Some projects have earned criticism from 'old-guard' communists: factories have been known to use capital to build hotels instead of replacing their old machinery, attracted by the higher expected profits. In similar fashion, township and city governments have used funds under their control to invest in the service sector and other activities, sometimes in other locations. Questions of ownership and control have become very muddled.

A crucial part of this change is a completely new attitude to how the state directs investment funds under its control. Instead of receiving all 'profits' from state-owned enterprises, it now charges taxes (which are lower and enable the enterprises to retain more of their 'profits'). These are then recycled through the banking system, instead of by central allocation of invesment funds. The banks are expected to supervise loans, and the state to manipulate this new credit system by means of indirect controls such as the rate of interest.

Unfortunately, the banks have little experience in the administration of lending; factories have borrowed to pay workers bonuses, or for projects whose merits or viability the bank is unable to judge. However, once such experience has built up, in spatial terms this shift in funding is likely to favour enterprises in areas which have a higher certainty of success and repayment. This is likely to reinforce the dominance of the Coastal Region.

Lower-level local authorities

One of the boom areas of industrial and commercial development during the reforms has been in small- and medium-scale enterprises. These have been initiated at the level of county, township and smaller cities, often by the local government alone or in collaboration with private entrepreneurs. Though technically called collectives, these businesses are often under the control of individuals, and it is frequently local CPC leaders who have used their power to get in on the act (of decentralisation) early.

It is such locally initiated industry and commerce which has accounted for a large porportion of the rapid economic growth of the reform period. But such development has also contributed to the 'overheating' and inflation problems of the economy, because of difficulties in constraining levels of investment and ability to bid up prices for raw materials and components, as well as the demand for construction materials used to build them.

There is a significant regional aspect too: this industrial and commercial growth is happening above all in parts of some coastal provinces. Several factors contribute to this 'bias'. Such areas may enjoy sub-contracting relations with existing nearby large-city industries; the purchasing power of the population is higher so new products can sell; in some cases, foreign capital or investment from overseas Chinese back in their home area is significant; and surplus peasant labour in rich agricultural areas can be employed using rural-generated capital. This is evident in parts of peninsular Shandong. In one county town, where party leaders have set up a range of new industries, 2,000 out-of-towners are employed and accommodated in dormitories.

The example of south Jiangsu is also very significant, as described in Case Studies 6.1 and 8.1 on pages 160 and 220 respectively. Guangdong province demonstrates the most rapid growth of this type involving foreign funds. But the leader overall has been Zhejiang, which, like Jiangsu, abuts Shanghai. Even in the earlier years of the reforms, Zhejiang saw much more rapid growth in the collective than in the state-run sector, so that by 1985 it was producing far more in the collectives, way ahead of other provinces.

Provinces

The central planning system was based largely on Beijing's control of enterprises and agriculture through the provinces and even through lower layers of local government; in effect there was a 'dual-control' system. Provinces have therefore been a key spatial as well as political unit in the implementation of state economic policy. The economic reforms have shifted this role considerably, weakening the planning function of provinces as agents of the centre.

The intention of the urban reforms was partly to reduce the rigidity of the planning process, which treated provinces as more or less sealed boxes acting as local components of the plan. Whereas before there was little chance for co-operation between enterprises on different sides of a provincial border, they were now permitted to find their own suppliers of raw material and components, and to sell their own output. This should have increased efficiency by encouraging spatial concentration and more competition. In some respects this has worked, and where enterprises have been able to escape the constraints of the former plan the province has become less significant as an economic unit.

But in other ways provinces have seized on their greater autonomy to intervene in new forms of activity, and, in effect, to build up their own power in other ways. These include using protectionist measures to benefit enterprises under their control. Thus the more market-oriented intention of the last decade of reforms has been subverted by decentralisation. New problems have been unleashed which alter the economic relations between provinces. The result runs counter to the intended increase in trade and investment across provincial borders. For instance, some province leaders have taken the opportunity to develop industries to produce goods locally which before they had to 'import' from other provinces. To protect them from the rivalry of the previous suppliers, tariff barriers have been put up to keep others' goods out.

A further twist to this is that these new 'infant industries' may use local raw material supplies, which are then denied to traditional users who might well be more 'efficient' producers. This situation has arisen in a number of inland provinces, creating conflict with the normal producers in the Coastal Region. Thus attempts to increase the 'value-added' produced within their boundaries, rather than act as mere raw material suppliers to others who gain those income benefits, are a constraint on the intended role of the coast. An example is Gansu province in the west, an important wool producer which previously supplied this raw material to long-established carpet factories on the coast. Now it has become the second largest producer of carpets in the country. The desire of the government to promote regional specialisation is neatly subverted.

Such developments have led to charges that provincial leaders are acting like 'economic warlords', a term which recalls the collapse of China in the 1920s and 1930s into separate mini-states ruled by local warlords. There are numerous examples: one industry which is particularly significant is silk textiles. Because most of the output is exported, whoever controls this can earn foreign exchange and use it (or abuse it) to purchase imports, or sell it above the official rate in China.

In Shanghai, normally the main processing city for silk goods, the industry had almost shut down in 1988. It was being starved of raw silk because producing areas were keeping most of it for their own new factories, which increased their wealth and enabled them to earn foreign exchange. So determined were some provincial authorities to ensure supplies for their own factories that they set up armed border guards to prevent silk leaving.

By the late 1980s the central government was trying to prevent this erosion of its control over foreign exchange and introduced a tax on exported silk goods at 100 per cent. The intention was to force others to export through the state monopoly corporation, which then receives a complete rebate on this tax. This competition for raw materials has affected other industries too, leading to inter-provincial conflict and regional rivalries of a new type.

In an attempt to control the local economy, provinces are overstepping their legal authority, and the centre seems powerless itself to do much about it. In some ways central government embarrassment at what has been going on between provinces is at odds with its desire to promote a market economy. The problem it has not reckoned with is that markets encourage participants to compete by using any initial advantage they have, and this is being used for all it is worth. A situation is arising which may be worse than that under the central plans which market forces were meant to replace: a combination of the worst aspects of both market and remnants of central regulation.

Beggar-thy-neighbour attitudes are even more apparent in the actions of government and entrepreneurs in Guangdong province, which is part of the Coastal Region (Hong Kong is adjacent to it). The inland provinces have the excuse that they are trying to retain some control of income-earning opportunities in the face of the coast's unfair advantages. Guangdong, the epitome of coastal 'success', has no such defence. Its business practices have created conflict with neighbouring provinces and considerable resentment in Beijing and among other Chinese.

Inland provinces have argued that they should share in some of the experimental privileges enjoyed by Guangdong and the coast, since the Coastal Region is not simply earning through new business, but is distorting existing production patterns and weakening the inland provinces. Now Hunan province, for example, is being allowed to retain more of the foreign exchange it earns.

The most serious grievances levelled at Guangdong include those arising from its practice of buying raw materials or even finished goods from neighbouring provinces with Chinese currency (Renminbi), then selling them abroad so that the foreign currency earned never gets to the inland provinces. Because the foreign exchange can be sold to those who will give more yuan than the official rate, Guangdong entrepreneurs can then afford to buy from their suppliers at higher rates, so cornering the market (and boosting inflation).

This has provoked the authorities in neighbouring inland Guangxi,

Hunan, Jiangxi and even coastal Fujian to demand payments in foreign exchange, or to set up border controls on traffic in certain goods. Some inland provinces have tried to get in on the act by setting up businesses in Guangzhou (Canton) or the Special Economic Zones (SEZs), all of which are on Guangdong's coast (see Chapter nine).

Beijing's attempts to institute controls over some of the more flagrant violations have met with stiff resistance from local officials and businesses. In the late 1980s it appears that the central authorities made a real effort to implement controls over some of the more blatant private and official dealings in Guangdong. These included the cancelling of over-budget investments and even some joint ventures with foreign partners. But some leaders are adamant that their province should enjoy the fruits of the 'experimental' open status which had supposedly been granted to increase foreign investment and improve technology. They are aided by the fact that when the central government attempts to institute proper regulation foreign business confidence is reduced; it is difficult to control economic growth without constricting supplies of outside capital and technology.

Pressures for new province-level authorities

There have been at least two attempts to redraw province boundaries in the last five years, and this may be indicative of the new advantage in controlling a province. One was for the creation of a new province centred on the proposed huge Sanxia (Three Gorges) hydro-electic power (HEP) dam and flood control project at the Yangzi gorges. This has been shelved, and it is still uncertain whether the dam will go ahead because of cost and environmental protests. The other proposal was to separate the south-coast island of Hainan from its parent, Guangdong province, creating a new province; this was agreed in 1988.

The Hainan case is remarkable, because it will restore powers of trade and business to the island's authorities only a few years after a major scandal. In 1985 many party leaders on Hainan were forced to resign after they misused their autonomous trade powers by importing large numbers of cars and motorbikes from Japan which were then resold on the mainland in conflict with import controls on such luxury items.

The State Economic Commission's action in allowing larger cities (initially nine) to become, in effect, provincial-level authorities, at least as far as their economic responsibilities are concerned, has further increased the number of province-level units. This indicates how decentralisation has created significant advantages for provincial officials. They are able to reduce their role as agents in the regulatory functions of the centre, and increase their own control over existing local resources and the growth of new ones. Local power means being able to retain more control over the increased wealth and foreign exchange generated (some of which may find its way into personal benefits).

Impact of the 'open door' policy on location of economic activity

One of the earliest 'reforms' of the post-Mao period was the 'open door' policy, which greatly expanded China's economic connections with the rest of the world (see Chapter nine). From the low levels of the mid-1970s, trade had increased three times by the early 1980s. More significant than trade was the acceptance of foreign investment and borrowing of capital from foreign governments and banks. To the Maoists such policies were a sign of capitalism and subjection of the country to imperialism. Now China is a member of the World Bank and the International Monetary Fund (IMF), and even invests money abroad itself, for instance in an iron ore mine in Australia.

The spatial impact of the new trade and investment opportunities has reinforced the growth of the Coastal Region. The government's faith in the region's comparative advantage has also been backed with a number of policies. Of these the best known is the Special Economic Zones (SEZs), set up at four places in south China in 1980. In addition, and more significant, are the fourteen port cities and one area (Hainan island) which were later granted particular rights aimed at encouraging foreign investment. The entire island of Hainan was then also made an SEZ in 1988.

The central government has itself committed disproportionately high levels of capital to the improvement of transport and dock facilities in all these places, partly to encourage investment but also to aid in shipping China's own exports from the coast and inland.

The rapid growth of the Coastal Region, and the concentration in it of both enterprises and population with higher spending power, has attracted to it higher levels of imports of consumer and producer goods. In turn, the new emphasis of the urban reforms on light industry has been concentrated on the coast, and the central government is intent on raising the export capacity, too, in order to help pay for the country's imports. In the second half of the 1980s the import bill was far in excess of the value of China's exports: there arose a serious balance of payments problem.

There have been hopes that oil would be found offshore at several sites on the east and south coasts, but so far there has been little success. Oil is already exported in large amounts from the northeast to help pay for technology imports, and new supplies would have helped reduce domestic energy shortages and would have also increased exports. The exploration efforts have in the meantime further boosted the Coastal Region's growth (see Chapter nine).

Emergence of the Coastal Region as dominant

A whole series of factors, then, seem to coalesce to create coastal pre-eminence, and although some of its growth may be the product of comparative advantage, it is also a result of government policy preferences, and the way in which local authorities have used their lead over other provinces. It is difficult to separate out the causes of the relative success of the coast and decide how much to attribute to inherent locational advantage, to higher productivity, to greater enterprise spirit, and to government policy (e.g. the promotion of Special Economic Zones and open ports, or the improvement of infrastructure).

What is clear is that it is achieving investment which is higher than the other macro-regions in relation to its share of population (see Table 2.1). Taking account of all sources of capital, the investment in fixed assets in the Coastal Region in 1985 was 52.7 per cent of the national total, while the population represents only 41.3 per cent.

The economic power of some coastal provinces to control supplies of raw materials has provoked conflict with inland authorities, as we have already seen. This seems to have prompted prominent government leaders to modify the three-fold 'division of labour' of the macro-regions (see Table 2.2). In the late 1980s, Party Secretary Zhao Ziyang and others, while still promoting the Coastal Region's rapid-growth policy, were saying that the coast had to look for alternative supplies of raw materials, even those from abroad, so as to reduce the constraints that its demand has put on inland development.

They added that those industries using large quantities of raw materials or energy should be cut back, and stress should be put on the growth of lighter industry. The products should be aimed especially at export markets, so as to reduce the impact of selling goods inland in competition with other producers there. Clearly, a regional division of labour is producing its own political price, and the economic objectives of increased efficiency are not to be achieved without the emergence of social and political conflicts which themselves are unmeasurable but are real costs of such policies.

The east–west divide

The conflicts with the Coastal Region experienced by the inland provinces are most intensely felt by those in the Western region. The issues are made more complex by the fact that this region is inhabited by many people of non-Han 'minority nationalities', who have always been at the thin end of economic policies, which are more geared to benefit the dominant Han people (see Chapter eleven). This has led to the concept of the 'east–west' divide in China, which is in some ways comparable with the arguments about a north–south divide in Britain. Despite the fact that growth rates in the west are quite respectable, leaders and some people in the west feel that their region is being left behind, or even deliberately neglected, under state policies which so clearly promote the coast (see Table 2.3 for a comparison of regional growth rates).

Government assurances that the Western Region will become the major growth zone of the next century are little consolation for people who have to endure current official policy of merely developing farming and animal husbandry, even if some (including sections of minority peoples) in the Western Region have experienced increased incomes under the new commercial policies (see Figure 2.2). They resent the limited role assigned to them of supplying food, raw materials and energy, mainly for 'export' (on improved transport links) to the rest of the country. Central governments explanations that the thrust to develop the coast first is intended to provide China with the wealth to develop the whole of the country is probably little consolation.

But the grievances arise not so much from national minority groups such as the Uighurs, Tibetans or Mongolians, as from Han Chinese who have settled in the west in large numbers in recent decades. They are there often as a result of some compulsion, and resent being excluded from the benefits going to provinces in the east from which they may have been sent. So the east–west divide is a point of conflict for the west's leaders, who want to retain the central subsidies they are accustomed to, and people (mostly Han) who resent the place and function they have been given in the new policies.

The minority peoples have other reasons to be discontented with central government, and in some respects their experience of being ruled by outsiders has been aggravated under the reform policies. Indeed many of the non-Han people (as seen in protests in Tibet and Xinjiang) would prefer the Han to withdraw and leave them with greater independence from what they experience as colonial control.

The ethnic conflict arises largely out of the minorities feeling that the Han want to use their areas mainly for the benefit of the Han economy rather than theirs, which are in most cases based on different resources and systems of production. The CPC's commitment to equal rights falls far short in minority people's experience. Without really having any say, they see their pasture land taken over by Han settlers who cultivate crops for Han people, mines opened to provide the east with raw materials, territory used as a dumping ground for prison labour camps, and dangerous nuclear tests (in Xinjiang).

So although some of the indigenous people have benefited from greater commercialisation of their areas under the impact of the reforms, much of the thrust of the new policies is seen as intens exploitation of their areas, rather than as sympathetic to nee are able to define for themselves.

Not all of the Western Region is affected by this ethnic divide: the most populous provinces in it – Sichuan, Shaanxi and Guizhou – are predominantly Han. They seem to have been included in the Western Region in order to help promote their neighbours' growth. They were also significant recipients of investment in the *sanxian* policy, and this industrial and scientific base is intended to provide a lot of ready-made facilities which just need to be 'civilianised'. This is not very easy, given their remote locations with long routes to markets in which they have to compete with much better-placed enterprises. Being expected to compete on this unfair basis is another element of the conflict creating the east–west divide, to which the central government has said it will pay more attention.

Policies for the west and minorities

Whether this attention is much more than lip-service remains in question, since the central government seems to have neither the will nor the ability to hold back the coast. However, it has been necessary to go some way to recognising the east–west gap, and formulate policies for dealing with it. The 'east–west dialogue' is an example: an attempt from the early 1980s to encourage enterprises from the east to participate in joint ventures or direct investment in the west. Examples of such projects in Xinjiang include fruit canning and other food processing, a truck assembly plant, and a textile factory.

The policy's success has been very limited, with only a small increase in value of output from such ventures. There was a more concerted effort to promote a large number of projects in Tibet, the least populous of all the Western Region's provincial-level authorities, in the mid-1980s. These were mostly in the service sector and infrastructure, and were criticised in Tibet because many of the forty-three projects seemed to be aimed mainly at increasing the capacity for tourism, the dollar earnings from which would benefit the central government.

By the late 1980s other ideas were being promoted. These included increasing minority area trade contacts with the nations bordering them, so that they 'look in two directions simultaneously', thus 'converting minority regions from remote places far from domestic markets into frontier areas adjacent to an international market' (*Beijing Review* 27 March 1989). Cross-border contact is unlikely to produce much in the way of growth, but in terms of morale it is probably quite important, especially in areas where people are ethnically the same on both sides of the border and want improved contact. The most benefit from this is likely to go to Xinjiang after the opening of the new rail link from Urumqi to the Soviet Union and thence on to Europe.

Another proposal is to expand the capacity of the minority regions to exploit mineral and other resources, using 'continued grants of intellectual, material and financial assistance from the state' (*Beijing Review* 27 March 1989). This reference to 'continued grants' is a diplomatic response to the anger expressed by leaders of western, provincial-level authorities; in 1987 the state announced that central subsidies were going to be phased out, leaving deficit provinces (like most of those in the west) to cover their budgets in other ways.

This obligation comes at a time when many of the west's skilled people are trying to leave for the east, reducing the development capacity of the region. In general, Han people do not like to live in national minority areas, and, since the 1950s, government posts have attracted salary bonuses to encourage people to go and work there. But with greater freedom of movement for people in the reform period many skilled workers, scientists and other 'intellectual resources' have been leaving the west in search of a better life further east.

The unimpeded impact of the reforms on the west is likely to further antagonise most of the minority peoples, who see the new policies as designed for the benefit of others. However, leaders in the Western region are in an awkward situation; to remain leaders they have to support the reforms (even if they are from the ethnic groups), yet at the same time be heard to complain of the reforms' unfairness and harmful impact on minority nationlities. The sort of protest movements seen in Tibet and Xinjiang in recent years, complaining that the local governments are inadequately expressing the opposition of the minorities to the impact of the Han reforms, may well continue.

Economic crisis and the reforms

Implicit in the attitude of letting some areas get rich 'first' is an official faith in the 'trickle-down' theory of economic development, in which areas growing first and fastest stimulate poorer areas nearby through their increased purchases of goods and raw materials, hiring of labour, and eventual cross-investment. In my view, it is largely discredited in the eyes of many people in less-developed countries.

In practice, it can allow rapid economic *growth* – as has indeed been happening in China in the last decade – but is much less effective in the promotion of medium- and long-term income distribution and welfare provision. Some of the most serious economic and political problems facing the leadership in the late 1980s relate precisely to this inadequacy. 'Overheating' of the economy and inflation at levels between 20 and 30 per cent produced widespread disenchantment with the reforms, and led to a drop in real income for many urban inhabitants.

Another spatial aspect of the regional divergence is the extra strain it puts on already overloaded transport and energy sectors (see Chapter seven). Since the 1950s there has been a serious shortfall in transport and energy provision for the economy and domestic use. There is still no effective plan to increase their provision at a rate which even catches up with the pre-existing gap, let alone matches the economy's growth. Interestingly, part of the 'comparative advantage' proposals for the Central Region is to improve the use of its energy resources and to promote the movement of energy-hungry enterprises from the coast to the centre. This would help reduce demands on the transport system, much of which is used to shift coal to places where electricity can be generated without long transmission lines.

REGIONAL POLICY IN THE 1980s

Regional policy has emerged during the 1980s in parallel with the reforms, and is largely concerned with improving the production conditions of the Coastal Region. Since this region is already in advance of the rest, living conditions of its people are likely to improve further. As well as the higher levels of investment going to the coast, government policy favours this region in the concessions granted for foreign capital. The coast has the five SEZs, as well as the fourteen coastal cities and other coastal 'open zones' with similar 'open door' freedoms and more infrastructure spending from the government (see Figure 9.4 on p. 237).

The idea of regional policy is mainly to strengthen the economy of the already better-off coastal provinces. By contrast, in Western countries it has generally been the objective of regional aid to improve living standards in worse-off regions by encouraging more economic activity there, trying to achieve a more equitable spatial distribution of employment levels, income and welfare facilities.

China's leaders argue instead that a stronger, economically developed and more technically advanced Coastal Region will provide a better basis for future assistance to the poorer parts of the country. The coincidence here with Mao's view in his 1956 speech ('On the Ten Major Relationships') shows how difficult it is to unravel politics and economic policy in China, given the usual simplistic presentation by whichever faction is in power of their total opposition to whoever went before them.

But apart from the major support for the coast as a macro-region, policy in the 1980s has also promoted some new types of smaller regions, some of which encourage not only inter-province linkages, but also connections between the three macro-regions.

Regions in the Seventh Five-Year Plan

One of the most interesting aspects of the Seventh FYP (1986–90) for geographers is the stress it puts on regional differences, and the proposals in it for different sorts of regional policies. In fact, new types of policy region (i.e. regions designed to achieve certain objectives) seem the vogue in China. Apart from those in the Seventh FYP itself, a confusing welter of new regions have been proposed (but not all implemented) by different government institutions and other bodies. They often seem at cross-purposes and pull in conflicting directions.

These new types of region are one of the most significant agents in the dissolving of traditions of provincial power-broking. They are aimed at improving efficiency by encouraging enterprises and local authorities to deal directly across province boundaries. In the old rigid planning system, managers had to deal up the bureaucratic hierarchy to the province authorities and down again to communicate with their neighbouring plants. Production connections between provinces were discouraged under this system, since the provinces were in effect closed systems which dealt with Beijing as their superior but not with equal-level neighbours.

In marked contrast, the new regions which were sanctioned by the Seventh FYP (some were already in operation) aimed at reducing the significance of provinces as economic managers. Instead they are intended to promote economic efficiency by encouraging links between enterprises no matter which province they are in. The emphasis is on promoting horizontal linkages, against the 'verticalism' of central planning. Some of the largest new regions have been backed up officially by high-level support, including the appointment of government ministers responsible for them.

At this scale, there are different types of region aimed at distinct problems or objectives. For instance, in an attempt to improve energy production and use there is the Shanxi Energy Base region, centred on that province but linked also to parts of Inner Mongolia, Hebei and Shaanxi. (It falls in the Central Region, which is meant to focus on energy production anyway.)

There are three other large 'horizontal' regions on this sort of scale. The Shanghai Economic Zone, and the Northeastern Economic Zone aim to improve links between adjacent provinces which already have large amounts of industry. Both include provinces from the Coastal and Central Regions. The former incorporates Jiangsu, Zhejiang, Fujian, Anhui and Jiangxi, with its headquarters in Shanghai. The Northeastern includes Liaoning, Jilin and Heilongjiang.

The remaining one connects three provinces from the Western Region – Sichuan, Yunnan and Guizhou – and Guangxi Autonomous Region, which is part of the Coastal Region, despite being one of the country's poorest provinces. It appears to have the transformation of the *sanxian* production capacity as its focus (see Figure 2.4).

These large regions are the first level of a hierarchy of the new policy regions, which are described as 'networks of economic zones'. This language itself indicates how far the thinking has gone away from the

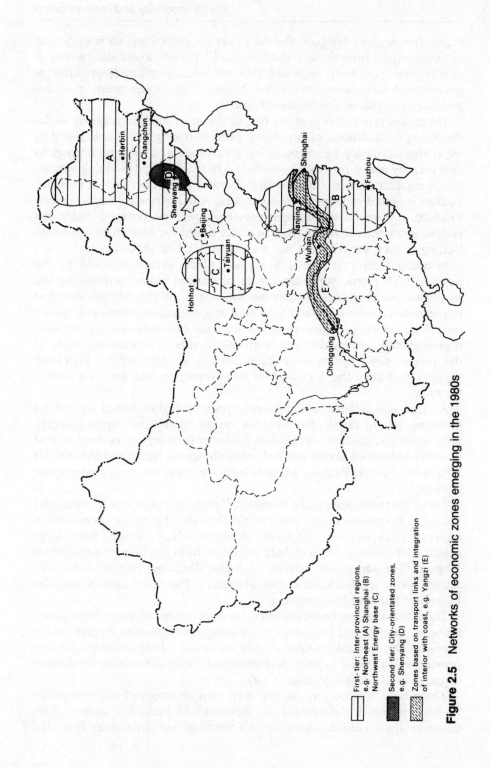

Figure 2.5 Networks of economic zones emerging in the 1980s

First- tier: Inter-provincial regions,
e.g. Northeast (A) Shanghai (B)
Northwest Energy base (C)

Second tier: City-orientated zones,
e.g. Shenyang (D)

Zones based on transport links and integration
of interior with coast, e.g. Yangzi (E)

A

Harbin

Changchun

Shenyang

Beijing

Taiyuan

Hohhot

Nanjing

Shanghai

Wuhan

Fuzhou

Chongqing

B

C

D

E

notion of provinces acting as the state's agents in a strict vertical planning system. In the next level are found networks 'along vital communication lines'. So far, two seem to have been mentioned, based on the Chang Jiang (Yangzi) River and on the crucial rail route that links the east coast with Urumqi in the far-west Xinjiang Autonomous Region (soon to connect China through the USSR with Europe). These curious 'linear' regions (see Figure 2.5), whose organisation has not been clearly spelled out, cross all three macro-regions. Really they are not regions as such, but rather a co-ordinating conference of the relevant city authorities.

The third level of new region types link together groups of cities into smaller regions, apparently to create logical cross-border connections where there are industrialised areas straddling provincial boundaries. Such groupings seem to have been springing up partly as a result of provincial initiatives, in recognition of the mutual benefits to be gained. There are examples within the northeast, in a grouping around Shenyang, and in the grouping of Tianjin and cities in Liaoning, Hebei, and Shandong around the 'Bohai rim' (the coast of the Gulf of Bohai). It is especially this type of region that seems to be at cross-purposes with the others. Given that these smaller ones seem to contain much of the relevant productive activity, higher-level efforts would have a rather contradictory impact on them.

PROSPECTS FOR THE COMING YEARS

The most significant spatial impact of the economic reforms is in the reinforcement of the strength of the Coastal Region, and the stirring of inter-regional rivalries and conflicts arising from the resultant 'east–west gap'. This major regional advance for the coast is a product of both the working of the economic reforms (which promote the comparative advantage of that region) and deliberate regional policy. It is reinforced by the powers which provinces have taken for themselves in the decentralisa-tion of planning, and by the formation of the new types of region which aim to cut across provinces but which are also most numerous in the Coastal Region.

In short, there is unlikely to be any reduction in the existing strength of the coast and its potential for rapid growth, unless the conflicts arising lead to a significant shift in official policies. Even such a shift is of doubtful value, as the central government may be too weak to rein-in the economic-ally strong coastal provincial governments and other business interests.

Aspects of regional policy designed to reduce spatial inequity are relatively weak, though there is some evidence of growing concern about increasing spatial (including regional) inequality. There are some aid provisions for poor areas, and there is greater willingness to reveal the level of rural poverty in mountainous and hilly areas. Paradoxically, such areas of poverty occur even within coast provinces, reminding us of how crude are the macro-regions in characterising their internal coherence.

The most crucial determinants of poverty in China are poor farming conditions (of soil, slopes, access to water) and the consequent inability to produce surplus for sale on the market, or of poor transport access to markets even when a surplus can be produced. These are conditions which are largely a factor of economic and regional policy, and it is going to be extremely expensive to overcome them. It is worth noting that under the reforms the number of counties eliglible for special poverty aid has doubled, though the fund for it has not been increased. Polarisation between richer and poor areas is likely to continue and become more acute, producing greater social stresses.

One way in which people are responding to the inequalities of earning opportunities is by migrating, sometimes seasonally, in response to agricultural work, but increasingly in search of permanent employment. This process is now possible, whereas during the Maoist period place of residence was much more strictly controlled. Migrants can buy food on the open market rather than use ration cards which are given to urban dwellers through their place of work. Though a market in labour has yet to be officially sanctioned it is developing by default and is not being strictly controlled. This is likely to continue, as central and local controls on movement are extremely difficult to re-impose without harmful effects on other aspects of the economic reforms (see Chapter eight and Case Study 5.1 on p. 126).

Lack of data makes it is difficult to assess the patterns, but it is likely that movement correlates strongly with economic opportunity, so that growth areas in coastal provinces are major recipients. It has been reported in the *People's Daily*, for instance, that there are 2.5 million people in Guangdong from neighbouring provinces either in or seeking work. After the New Year festival in 1989, in the capital Guangzhou (Canton) alone there were 30,000 rural labourers waiting in a main square for temporary jobs. By comparison, the size of the workforce in the province in 1985 was around thirty million.

While work in the towns and cities is far from guaranteed under such conditions it is unlikely that agriculture will ever again be able to absorb more people. Even if work is not easily available, the mobility of labour under the new conditions is likely to continue, with increasing pressures of urbanisation and residence in towns and cities. Stronger regionalisation of economic growth and the emergence of patterns of smaller towns in the periphery of larger cities along the model of south Jiangsu seems likely (see Chapter eight).

China seems set on a course which is going to produce increased regional differences and heightened political awareness of the problems such spatial inequalities involve. In the pursuit of economic growth rather than a broader, more equally shared development, Deng Xiaoping's model aims at producing greater wealth and welfare for all in the long term even if it means greater inequality in the meantime. His objective is shared by many, but not all, of the central leadership and is increasingly opposed by significant sections of the people.

But the forces unleashed by the commercialisation of socialism have created powerful regional and provincial interest groups in better-off areas which desire to maintain the decentralised and market-oriented policies. With the centre much less able to reinstitute controls there is likely to be a growth in 'separatist' aspirations by the well-off and growing tension in the less prosperous areas. Some poor hill and mountain areas have already suffered food shortages, and because aid is inadequate and political will insufficient they are potentially victims of the new spatial patterning of Chinese society, voiceless and superfluous to central or local needs.

ACKNOWLEDGEMENTS

My thanks to Roger Chan, Janet Hadley, Maurice Howard, Richard Kirkby and Martin Lockett for their comments and help in writing this chapter.

REFERENCES

Buchanan, K. (1970) *The Transformation of the Chinese Earth,* London: Bell.
Naughton, B. (1988) 'The third front: defence industrialisation in the Chinese interior', *The China Quarterly* 115, September: 351–86.

FURTHER READING

Geography (1989) October.
   The annual 'This changing world' supplement in this issue is devoted to China on the fortieth anniversary of the People's Republic and contains short articles of relevance to regional and spatial issues. Also published as a separate booklet available from the Geographical Association.
Goodman, D.S.G. (ed.) (1989) *China's Regional Development*, London: Routledge/RIIA.
   This collection of papers includes coverage of regional policy, the problem sectors of the economy (energy and transport), the impact of foreign trade on regionalisation, and case studies of the Special Economic Zones, a province from the Coastal Region (Shandong), the Central Region (Shaanxi), and the Western Region in general.
International Regional Science Review (1987) 11, 1.
   A special issue of this journal devoted to 'Regional aspects of the Chinese economic reforms', containing essays, almost all of which are relevant and some excellent.
Leung, Chi-Keung and Chai, J. C. H. (eds) (1985) *Development and Distribution in China*, Hong Kong: University of Hong Kong.
   Some useful contributions relevant to regional and spatial issues.
Naughton, B. (1988) 'The third front: defence industrialisation in the Chinese interior', *The China Quarterly* 115, September: 351–86.
   A rare study of the *sanxian* policy and its regional impact.

UPDATE

The official Chinese magazines *Beijing Review* and *China Reconstructs* (now renamed *China Today*) often have articles or news of matters concerning regional development. The Hong Kong weekly *Far Eastern Economic Review* carries reports and analysis which often concerns particular provinces or regions, and is a very useful counterbalance to the bias of official sources. More difficult to get hold of, the Hong Kong newsletter *China News Analysis* and the Japanese trade monitor *JETRO China Newsletter* are well worth checking for their occasional well-informed articles on regional development and policy (JETRO is the Japan Export Trade Research Organisation).

Chapter three

History: historical perspectives on the current geography of China

Richard Louis Edmonds

To understand contemporary China it is necessary to know something of the country's past. In this chapter we explore the role of history in certain aspects of China's geography, and mention a few ways in which the past has resurfaced since 1978.

Governments in China have had a profound impact on the country's geography. Moreover, the continuity of history is still to be seen in its present government, despite its being generally considered radically modified. After two militarily successful revolutions in the twentieth century, as well as attempts at bureaucratic reform since 1978, the form of government owes much to the nation's dynastic past. Imperial-style bureaucracy is evident even in the present day. Athough China has adopted republican terminology since 1911, the state today is ruled by individuals who appear to hold power for life. This is more in the imperial than the republican tradition.

Under the Chinese emperor there existed a large centralised bureaucracy whose qualifications for service were based on the results of state examinations or personal relations with the ruling faction. Today, university examinations and party membership are the bases for top jobs in areas of society that are under central-government control and located in the national capital.

Despite being in the process of compiling a thorough law code, there is a feeling among many inhabitants that law is enforced rather arbitrarily. This increases with distance from the national capital and major regional cities. That rule by law is more proper than rule by individuals is often stressed in the state-controlled press, and this indicates that rule by individuals rather than law is still widely practised. The concept of rule by men of virtue was expounded in the classics by Confucius, with interpretation of the law traditionally at the discretion of local officials.

Dates such as 1911 (founding of the Republic), 1949 (the People's Republic) and 1978 (the beginning of the open policy and the assumption of power by Deng Xiaoping) do not represent complete breaks in Chinese history but are merely times when formal modifications occurred in its politics (see the economic and political changes listed on pp. 24–7). Without a historical background we cannot easily understand what makes Chinese governments operate in the fashion they do. The same can be said about the landscape. With this in mind let us take a closer look at some aspects of China's geography from a historical viewpoint.

CONTEMPORARY GEOGRAPHY IN HISTORICAL PERSPECTIVE

China's human geography is a result of the country's long, continuous cultural history. Over the millennia distinctive cultural regions developed in East Asian civilisation. In contrast with Europe or pre-colonial South Asia, however, these regions rarely led to the formation of separate nation-states.

Regionalism

Within China's borders are found many closely related but mutually unintelligible languages of the Chinese language group (See Figure 3.1 and Case Study 3.1 on p. 78). Each of these has a regional focus and possesses many dialects.

In addition there are languages spoken by China's minorities which are completely unrelated to the Chinese language group, including those of peoples with distinct geographical and cultural identities such as the Tibetans and Uygurs, who have been only marginally controlled by the Han-Chinese at various times.

We may wonder why the many linguistic groups of mainstream Chinese (Han) did not split into various nation-states, like the Romance-language-speaking peoples of Europe. One of the many reasons why the distinctions of various Chinese languages never fostered feelings of regionalism is because the Chinese written language has been more or less unified for 2,000 years. Prior to the twentieth century, Classical Chinese (a written form which played the equivalent role of Latin in Europe or Sanskrit in South Asia) was the mainstay of inter-regional communication and was read by all literate Chinese. Despite great regional linguistic variety, national identity for the majority Han-Chinese was never in doubt.

Regional variation in China is not confined to language. When meeting for the first time, Chinese often ask each other 'Where (i.e. what province) are you from?' The answer given will lead the person asking the question to form all sorts of prejudgements about the other individual's habits, or even moral character. The frequency with which such a question is asked is far greater than in most English-speaking countries and reflects the continuing importance of regionalism in Chinese daily life.

In some areas such as Guangdong and Jiangsu provinces, regional identity is so strong that it is difficult for an outsider to penetrate local

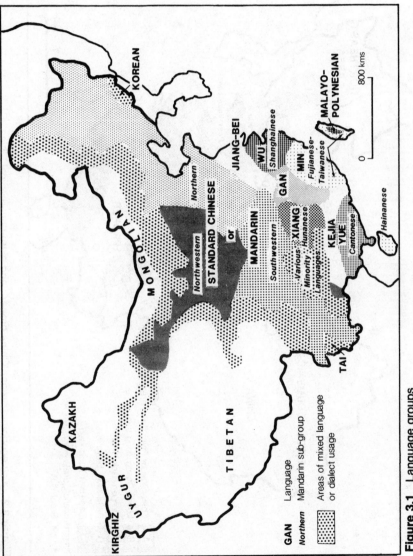

KIRGHIZ

KAZAKH

UYGUR

KOREAN

MONGOLIAN

Northwestern

Northern

STANDARD CHINESE
or
MANDARIN

JIANG-BEI

WU
Shanghainese

GAN

MIN
Fujianese-Taiwanese

Southwestern

XIANG

Various Minority Languages
Hunanese

KEJIA

YUE
Cantonese

TIBETAN

TAI

Hainanese

MALAYO-POLYNESIAN

0   800 kms

GAN   Language

Northern   Mandarin sub-group

Areas of mixed language or dialect usage

Figure 3.1   Language groups

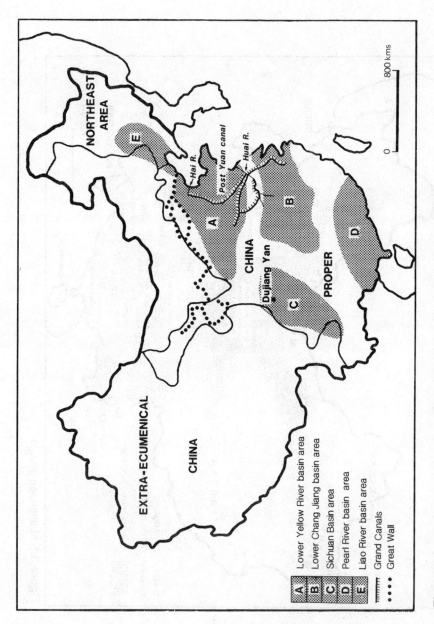

Figure 3.2 Key economic areas, 'China proper', and extra-eucumenical China

A Lower Yellow River basin area
B Lower Chang Jiang basin area
C Sichuan Basin area
D Pearl River basin area
E Liao River basin area
   Grand Canals
   Great Wall

NORTHEAST AREA

EXTRA-ECUMENICAL CHINA

CHINA PROPER

Dujiang Yan

Hai R.
Post Yuan canal
Huai R.

0   800 kms

society. Particularly along the southeast coast, but also in the interior, central authority has often been challenged and is still being challenged by regionally based groups. No doubt the mountainous terrain of southeast and southwest China, as well as the shortage of modern transport, plays a role in the persistence of these regional identities. At least since the Qing dynasty (1644–1911) the central governments have tried to break down regionalism by shifting officials and even peasants from province to province.

Although improved communication is likely to reduce Chinese regional distinctions to some degree, history and geography ensure that regional differences will continue to be significant for a long time.

Frontiers and history

On a larger scale, modern China can be divided into major regions (Figure 3.2). One is 'China Proper' (sometimes called 'Ecumenical China') which largely corresponds to the area where Han-Chinese have long been the majority. The other major region can be described as 'Extra-Ecumenical China', frontier regions where non-Han people are a significant proportion of the population. The northeast somewhat defies classification and therefore may best be considered as a yet a third major region. Liaoning province has long been settled by Han-Chinese, but has never been considered part of China proper. Jilin and Heilongjiang provinces to its north are now predominately Han-Chinese but were largely colonised by them in this century.

China has long had contact with surrounding peoples. Dynasties were established in China several times by non-Chinese people from Mongolia or the northeast (Table 3.1). The most famous of these were the Khitan, whose dynasty is called the Liao; the Jürched (Jin dynasty), who conquered northern China, and the Mongols (Yuan dynasty) and the Manchus (Qing dynasty), who subdued all of China. These 'barbarian' groups crossed into China from the north after learning administration and military organisation and technology from the Chinese. Likewise, during periods when China was strong, dynastic control extended beyond the Great Wall into the steppe.

The Great Wall was not one massive undertaking but initially a series of individual state's walls built during the Eastern Zhou dynasty. These were linked together for the first time under Qin Shi Huang (Ch'in Shih Huang-ti) during the third century BC. (He is famed for his elaborate burial arrangements, which include the terracotta army discovered near Xian in the 1970s.) Since then walls have been built or rebuilt at various locations along different portions of the semi-arid steppe in concordance with what Chinese dynasties could effectively control. How far the walls, and consequently Chinese power, could be extended into the grassland steppes

Table 3.1 The principal Chinese dynasties with approximate dates (Wade-Giles romanization is given in parentheses)

Shang	1766(?)–1122 or 1028 BC
Zhou (Chou)	
Western Zhou	1122 or 1028–771 BC
Eastern Zhou	770–221 BC
Qin (Ch'in)	221–207 BC
Han	
Former Han	206 BC–AD 8
Later Han	AD 25–220
Three Kingdoms	AD 221–264
Western Jin (Chin)	AD 265–311
Northern and Southern dynasties	AD 311–580
Sui	AD 589–617
Tang	AD 618–906
Five Dynasties and Ten Kingdoms	AD 907–959
Liao, Khitan (far north and northeast)	AD 947–1125
Song (Sung)	
Northern Song	AD 960–1126
Southern Song (south China)	AD 1127–1275
Jin (Chin or Kin), Jürched (north China)	AD 1115–1234
Yuan, Mongol	AD 1276–1367
Ming	AD 1368–1644
Qing (Ch'ing), Manchu	AD 1645–1911

was limited by the extent that agriculture could be effectively pursued in this environment, which becomes increasingly arid towards the northwest. In contrast, pastoral nomads could only penetrate into the more humid southeast by abandoning their nomadic herding lifestyle.

Despite these fluctuations, China's northern frontier can be described as relatively static, since there has not been tremendous expansion or contraction of the Chinese state over the millennia. In contrast, China's southern frontier has steadily expanded southward, with conquest and acculturation of indigenous peoples commonplace, and their territory being incorporated into China proper as the land was colonised for agriculture. At times, Chinese political influence has even extended beyond the country's contemporary southern borders into Vietnam and to a lesser degree into Laos, Cambodia, Thailand and Burma. Moreover, until the Mongol Yuan dynasty conquest of Yunnan province, it remained beyond Chinese control.

In earlier times, much of south China was the homeland of what are now China's southern minorities, some of whom are linked to neighbouring peoples of Southeast Asia. Even though these people did resist Chinese conquest from time to time, they never threatened the Chinese core the way the northern nomads did. This apprehension of northern frontier peoples has even affected the location of the national capital. Militarily strong governments have had a northern capital and greater contact with the northern frontier. Weaker states have often located the capital in the south, with conquest of that weaker dynasty eventually coming from the north.

China has also shown a continued interest in extending its influence to the south. Its traditional agriculture adjusts more easily to the tropics than to the boreal Siberian forests. The country today still has its back to the north. Fears of the Soviet Union and Japan are greater than any threat from the south. In part, this interest in Southeast Asia also stems from the extensive settlement of Chinese in the region during the phase of British, Dutch, French, American and Portuguese colonial influence. Although current global political and economic realities play a part in China's southward orientation, one cannot help but notice a continuation of traditional Chinese frontier policy (see Chapter eleven).

Dynastic cycles

When Chinese look back on their history they conceive of two great periods of territorial expansion and political stability, the Han dynasty and the Tang dynasty (Table 3.1). Although other dynasties are famous for cultural and other achievements, the Han and Tang stand out as particularly strong periods. Most Chinese refer to themselves as Han-Chinese (*Han-ren*), though the Cantonese (southernmost of the Han-Chinese) designate themselves as Tang-Chinese (*T'ong-yan*) because the extreme south of modern China was incorporated into the state during the Tang dynasty.

Chinese history was traditionally interpreted by the Chinese themselves in terms of dynastic cycles. In other words, Chinese history was viewed as repeating itself. The Han and Tang as well as other long, stable dynasties were followed by periods of disorder and the break-up of China into small states. Out of disorder a leader eventually arose who unified the country and imposed strong central authority on the exhausted nation. For instance, after the Han various dynasties ruled parts of China until Yang Jian reunited the country and established the Sui dynasty. The Sui set the scene for the long and prosperous Tang.

If one attempts to transfer this cyclical theory to more modern times the Qing dynasty can be seen as a great dynastic period which went into decline, followed by a period of disorder during which various warlords, the nationalists, the communists, and the Japanese all tried to establish control. Eventually the communists (led by Mao Zedong) gained 'the Mandate of Heaven' and established the People's Republic. In this simplistic way the theory fits. More importantly, many Chinese feel the theory fits. Although the dynastic cycle theory cannot be used to predict the timing of future events it is frequently adopted by the Chinese public to define their present condition.

Probably the most famous unifier of China was Qin Shi Huang, founder of the Qin dynasty, precursor of the Han dynasty. It was he who first reformed and standardised the Chinese language and, as previously mentioned, linked together the northern walls of various states to form the first Great Wall. Axle lengths were standardised, improving communications, weaponry was concentrated in the hands of his own soldiers, merchants and their offspring were forced to guard the frontiers, and 'undesirable' books were destroyed in an attempt at thought control.

Many Chinese believe that Qin Shi Huang and Mao Zedong are comparable. Under Mao's rule China was reunified, the Chinese written language was simplified, the railway gauge was standardised, the bourgeoisie lost its property, and writing was tightly controlled. Whether the dynastic-cycle theory is valid or not, the similarities between these two men are striking.

When asked about the authoritarian nature which men such as Qin Shi Huang, Mao Zedong and Chiang Kai-shek displayed, many Chinese reply that China is a *fengjian* (usually translated as 'feudal') society. *Fengjian* means different things to different people, but the idea that the state is ruled by one individual who is not to be questioned is the essence of this use of the term. Fortunately, men such as Deng Xiaoping and Chiang Ching-kuo (Chiang Kai-shek's son, who died in Taiwan in January 1988) have allowed more criticism than their immediate predecessors, Mao Zedong and Chiang Kai-shek. The system of one-man rule, however, remains unchanged in both mainland China and Taiwan.

Oriental despotism

China is notorious for floods and drought. In particular, the North China plain has flooded every other year for the past 3,000 years. The same variation in the monsoonal winds that can produce floods also can cause serious drought, especially in north China. Inevitably, governments have had to deal with these hydraulic problems. Karl Wittfogel in his book *Oriental Despotism* includes China within his category of 'hydraulic societies'. Wittfogel sees unpredictable climate and low level of property-based industrial development as opportunities for a state to develop despotic statecraft while constructing large-scale hydraulic projects.

Examples of magnificent hydraulic works in China include the Grand Canal, first constructed in the early seventh century AD to insure adequate grain supplies for what is now the Beijing area, and Dujiang Yan, a moveable weir built in conjunction with a massive irrigation project near Chengdu in Sichuan province by Li Bing during the Qin dynasty (Figure 3.2). Inability to maintain such projects is seen as one sign of a weakening central authority.

After warfare ceased in 1949, the new government of the People's Republic turned its attention to hydraulic works. In particular, many hydraulic projects were carried out in the Lower Yellow River (Huang He), Huai River and Hai River basins during the 1950s and 1960s. Despite criticism recently levelled at the Maoist period, especially for its authoritarianism, the efforts devoted to irrigation and flood control stand out as one of that regime's strong points.

Critics of the Deng Xiaoping regime say that the Chinese government is paying too little attention to the upkeep of hydraulic works. Canals and dams are said to have fallen into disrepair since 1978. Large-scale *corvée* (compulsory labour) is no longer popular. It remains to be seen whether flooding or drought will re-emerge as major problems in China. Clearly, seasonal water shortages and flooding in certain areas such as the North China plain, still cause disaster.

Key economic areas

As an overlay to the largely political, historical-oriented dynastic-cycle theory, in 1934 Chi Ch'ao-ting put forward an economic and geographical-oriented theory of key economic areas in Chinese history. Chi saw traditional China as divided into regions which were self-sustaining and independent of each other. In the absence of mechanised economic organisation, modern state centralism was impossible. The size and area of these regions changed from time to time but the areas roughly corresponded to provinces (Figure 3.2). The area of greatest agricultural productivity with facilities for grain transport was the key economic area for the conquest and unity of traditional China.

Chi placed his ideas within the context of the dynastic-cycle theory and Wittfogel's early work on the role of hydraulic projects in the politics of China. During 'the first period of unity and peace' (Qin and Han dynasties) the lower Yellow River and its tributaries was the key economic area. However, during 'the second period of unity and peace' (Sui and Tang dynasties) the lower Chang Jiang (Yangzi) River basin assumed the position of key economic area and was linked to the old key economic area, the lower Yellow River basin, via the Grand Canal. In the 'third period of unity and peace' (Yuan, Ming and Qing dynasties, AD 1276–1911) the key economic area remained in the lower Chang Jiang valley, although these dynasties were preoccupied trying to develop the Hai River basin as a new key economic area southeast of modern Beijing.

Chi stated that his theory was not applicable to post-1842 China because, due to the Opium War and other subsequent defeats China suffered at the hands of the European powers, the traditional Chinese world order had toppled after that date. However, if one were to use Chi's terminology, the early Republican period (1911–28) would represent another 'period of division and struggle' with the nationalists trying to re-establish an economic base in the lower Chang Jiang basin while various warlords and the communists attempted to set up rival bases. The establishment of the nationalists at Nanjing (1928–37) would be a transition stage which, of course, was disrupted by a Japanese (barbarian) invasion. The People's Republic could then represent the beginning of a fourth 'period of unity and peace'. As it is, one of the communist regime's early tasks was to construct hydraulic works on the Hai River, fulfilling the goal Chi noted for the dynasties of his 'third period of unity and peace'. However, the communist Chinese government has been able to use modern mechanised transport and communications to unite the old key economic areas, including Chi's secondary key areas (the Sichuan basin and the Pearl River basin), as well as effectively to control the whole country. In this sense, the old key economic areas have disappeared with the advance of modern communications. Yet with the addition of the Liao River basin, which was developed into a major industrial centre only in this century, Chi's economic areas equal China's modern industrial bases, and, as such, the key economic areas of traditional China leave their mark on the geography and politics of the nation.

The high-level equilibrium trap

High population density in China is not a new phenomenon. Centuries of relatively high population densities have had a profound impact on the country's geography. Downturns in population caused by epidemics, famine and warfare have occurred. Mark Elvin and others argue that the high density of population had trapped the country at a low level of technology by the mid-eighteenth century, with the beginnings of this process going back perhaps to the fourteenth century. China lacked source materials and had an over-supply of labour which forced it to rely on labour-intensive techniques rather than adopt mechanisation. Therefore, an equilibrium had been reached between resources and the level of technology employed.

The problems elicited by the Maoist regime's policy (perhaps forced upon that government by the USA and the USSR) to 'go it alone' and find indigenous answers to stimulate industrial development and increase crop yields after the Sino–Soviet split in the 1960s, combined with rapid population growth, might be viewed as a continuation of this high-level equilibrium trap. Under the Maoists, China was largely closed to external markets, only able to expand arable land very moderately, and lacked capital. The expanding population meant that the country continued to have a low level of productivity per capita. The resulting modest growth was due to peace, agricultural land consolidation, multicropping, and a modest increase in mechanisation. The nation's per capita yields remained low.

Since 1978 we have been witnessing another attempt to use foreign technology and capital to resolve the high person/land ratio problem, such as was tried by the Qing government in the late nineteenth century, by the nationalists in the 1930s, and by the communists in the early 1950s. The Qing attempt was internally aborted, the nationalists' plan never got off the ground, and the Soviet-inspired model of the 1950s was terminated by the Sino–Soviet split. The current policy can be seen as two-pronged, since the government is also trying to reduce person/land ratios by introducing population control.

Taiwan was able to move from an agricultural to an industrial economy during the 1960s and 1970s due to the island's infrastructure inherited from the Japanese colonial period, a heavy dose of American aid during the 1950s, the island's small size and relatively good natural conditions such as soil fertility and adequate precipitation, as well as the nationalists' consistent economic policy. Such an economic transition also occurred in Japan and South Korea. In each, development was not evenly dispersed geographically, and large movements of rural populations into cities where wealth accumulated more rapidly were the result.

It appears that the post-1978 communist government expects a similar pattern to occur on the mainland. Recent plans call for the eastern part of the nation to industrialise first, with the centre and west to follow. Therefore China looks to areas with geographical advantages which in the past had developed quicker to continue the same process in the future.

Urban geography

Urban centres in China probably first appeared on the North China plain during Shang times (Table 3.1). Wheatley suggested that these early urban forms were ceremonial centres which were transformed into compact cities during the Spring and Autumn period (722–481 BC). The city has been the centre and symbol of authority in Chinese society for millennia.

The traditional Chinese capital city was laid out on a grid pattern, with the centre of authority located in the centre or to the north of centre. Provincial and lower level capitals mimicked the pattern of the capital, with possible modifications due to terrain and pre-existing urban forms. The city would be surrounded with a wall, although settlement would often sprawl beyond the wall as the city grew. The dominance of this symmetrical plan demonstrates the importance of politics in the traditional Chinese city.

County seats were often evenly spaced across the land, although the distance between such administrative cities and towns generally increased in the less densely populated periphery. In contrast with these administrative centres was a newer type of city which largely evolved with the arrival of Western powers on the China coast (Figure 3.3). Many of them (e.g. Shanghai) were mere villages prior to the arrival of Westerners and the

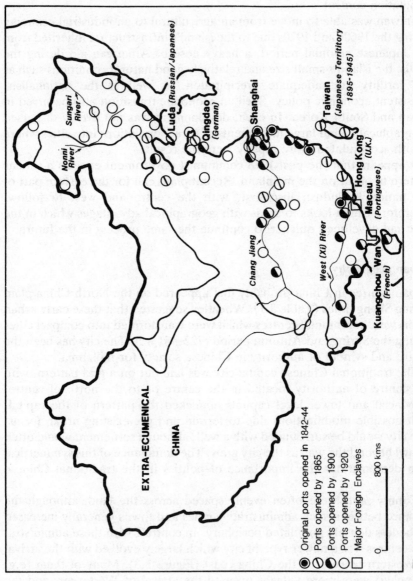

Figure 3.3 Treaty ports and foreign exclaves

granting of treaty-port or direct-colonial status. Non-administrative cities had existed in China prior to the sixteenth century. However, under Western influence the scale of such centres enlarged, and a regional focus began to develop. Most of the new cities were large ports located along the east coast.

In Maoist rhetoric this second type of city was formally considered 'semi-evil'. However, even under the Maoists, cities in which 'capitalism had sprouted', such as Shanghai, continued to grow. The traditional role of some major administrative cities such as Beijing was at the same time reinforced by the introduction of modern Soviet-style administrative buildings and monuments.

Since 1978 cities such as Beijing have continued to have their original grid pattern expanded and huge monumental government buildings are still being constructed. However, the economic emphasis of the new regime suggests that commercial functions, commercial cities and Special Economic Zones will get development priority. The physical difference between administrative and commercial cities, however, can still be seen in China today.

THE PAST'S INFLUENCE ON CHINA SINCE 1978

The open policy since 1978 has led to the reappearance of many past practices, so it is very important to be aware of China's traditions in order to understand recent geographical trends. In this section some of the more geographical aspects of post-1978 social change in China will be described.

Traditional burials

Shortly after the communists came to power in 1949, the party encouraged cremation as a practical way to save precious land and to dispose of the dead hygenically. People's mausoleums were established and open to all who did not belong to the 'landlord' class.

With the beginning of a more open policy in 1978 peasants again began to opt for ground burials. Government policy has offically criticised grave mounds as a waste of land, yet ground burials continue. These graves may not be the waste of land that the government claims them to be. Graveyards ensure that a plot of land will remain more or less covered with vegetation, which while not being economically productive, could be ecologically wise in the long term. It remains to be seen how much economic damage or ecological benefit grave sites will produce. However, they do embody the re-emergence of a traditional practice which many had thought was gone forever.

Traditional marriages

Since 1978 traditional marriage rituals have reappeared in rural China. Brides dressed in traditional red costumes with perhaps a few modern adaptations, and grooms coming to take their bride to the groom's family home can once again be seen in villages. There have even been reports of arranged marriages for children in some rural areas.

Although traditional marriage ceremonies and arranged marriages are very rare in the cities, old habits of conspicuous consumption at weddings are back, particularly among working-class people. Spending the equivalent of ten years' salary on a wedding feast and honeymoon is not uncommon. The impact of honeymoon travel on China's domestic tourist industry has been tremendous, with many of the trips covering great distances.

Religion

What Marx termed 'the opiate of the people' is tolerated in China today and is even encouraged in cases where it promotes tourism or foreign relations. The state is investing a considerable amount of money to restore Buddhist temples across the countryside and is staffing them with monks. These temples are open for tourists to visit, provided they pay a fee. As yet few young people are attracted to Buddhism, but many older people are praying at the temples again. Tibetan Lamaism survived the Cultural Revolution (1966–76) and is once again fervently practised among Tibetans.

Mosques in Islamic areas such as Ningxia, Gansu and Xinjiang now maintain contacts with Islamic groups in countries which have good relations with the People's Republic. In the last decade, several thousand Muslims from China have been on pilgrimages to Mecca.

Christianity in general has never had a large following. Catholicism is tolerated in the People's Republic of China in a nationalised form which does not recognise the authority of Rome. Since 1949 the Roman Catholic 'Bishop of Nanking' has resided in Taipei (capital of Taiwan) and the Holy See continues to recognise the Republic of China government in Taiwan.

Small-plot agriculture

Since 1978 individual farming of plots under the 'new responsibility system' (see Chapter six) give the land a physical appearance more like that found before the 1950s than the larger collective fields of the 1960s. The increased population dictates that individual field sizes must remain small. Larger fields have been divided using trees as markers, and one family may cultivate up to twelve separate plots. Consequently, much of the farm machinery formerly used on communes is now virtually useless in densely settled regions. In other areas (such as Jilin and Heilongjiang or the northwest) field sizes remain much larger. This was the general pattern prior to 1949. The crops grown have also changed since 1978, with a reversion to more varied cropping patterns.

Small-plot agriculture has brought with it the introduction of some Japanese- and Taiwanese-style techniques. Hand tractors are appearing in more prosperous areas and polyethylene sheeting is being used more and more to raise soil temperatures and to retain soil moisture. These techniques are not a return to the past but represent new technologies adapted to small-scale East-Asian-style farming.

Marketing

Much has been written on traditional Chinese rotating rural-market patterns. Itinerant merchants would visit markets in sequence, following the periodic market schedule. At the centre of these marketing regions there would sometimes be a larger market open on a daily basis. This network shared some geographical qualities with the central place system as described by Christaller and Lösch.

This type of rotating market was considered an aspect of capitalism and replaced with state-run stores in the larger market towns during the 1950s. At times, the peasant has been allowed a limited amount of free-market activity. After 1978 individual enterprise was again permitted in China, and by the early 1980s rural markets reappeared and expanded with a rotating pattern strikingly similar to pre-1949.

Art

Like the marketing system, traditional arts did not completely disappear after 1949, but new aspects of state-controlled art overshadowed those artists who desired to work with traditional themes. Socialist realism, borrowing much from the Soviet Union, became the state form of art. Art had a political message. Traditional art was forced into line as landscape paintings began to include hydro-electric power plants, communes, red flags and other aspects of socialist realism. Art is never meant merely to copy reality, and the socialist realism forms painted a picture of economic progress which exceeded reality in many cases.

Since 1978 the art propaganda policy has been gradually relaxed and artists are reaching out in two directions – to the West looking for new ideas and into the Chinese past. Of most interest to the geographer would be the revival of traditional landscape paintings. This may suggest a return on the part of some Chinese to a desire for greater harmony with nature rather than controlling and transforming nature for the benefit of humanity, which was the major message of Mao and socialist realism. Also, where monuments were destroyed during the Cultural Revolution, they are now being rebuilt or replaced by replicas.

Treaty ports

In the sixteenth century when the Portuguese arrived on the China coast a convenient arrangement was reached whereby Macau was set aside as a place where Portuguese could trade with China and where Portuguese law was to some extent in force. This was the beginning of the treaty-port era, whereby Western countries either claimed enclaves such as Macau, or established concessions under their control within Chinese port cities, such as occurred at Shanghai.

This treaty-port system was largely dismantled by the end of the Second World War. Recently, dates for the reversion of Hong Kong (1997) and Macau (1999) to Chinese control have been agreed upon, so that formal Western control of Chinese ports will come to an end. However, finalisation of the Hong Kong and Macau accords was eased tremendously by the knowledge that the 'open door' policy is producing open ports which bear some similarities to the old treaty-port system (see Chapter nine).

Today, the areas most open to foreign traders are along the east coast, where the majority of the treaty ports were located (Figure 3.3). Four Special Economic Zones (SEZs) have also been established at Shenzhen (next to Hong Kong), Zhuhai (next to Macau), and Shantou, all in Guangdong province, plus Xiamen (across the straits from Taiwan). In 1988 a new 'open' province of Hainan island was announced, and plans to expand the Pearl River delta economic region and the Chang Jiang delta region are also part of this coastal orientation. Although these locations are well placed to attract capital from Hong Kong, Macau and Taiwan, foreigners (both of non-Chinese and Chinese descent) are also encouraged to invest in these zones by the granting of special privileges (see Chapter nine).

The new system of open ports and Special Economic Zones differs from the old treaty ports in that the Chinese maintain all political, and the majority of economic, control. However, the coastal location and the idea of segregation of Chinese and foreigners in these enclaves is reminiscent of the treaty-port era as China and the foreign powers again try to find a way to meet on Chinese soil without conflict.

Private ownership

Private ownership is back in China as a result of the open policy. Of significance to the geographer, however, is the status of the last sacred cow of communal ownership – the land. Individuals can own the other means of production (tools, vehicles, farm equipment), and even if they let out a house their children can later inherit it. The land under the building, however, as well as the fields, still belongs to the state, though peasants are beginning to act as if they own the land which is contracted to them. Land ownership used to be crucial to a Chinese farmer's livelihood. Without a reversion to individual ownership we cannot say there has been a total reversion to past practices as regards private ownership, but certainly ownership of buildings, enterprises and other possessions has encouraged an awakening of people's traditional attitudes, customs and desires.

Whether the land ownership arrangement is merely a formality which will eventually fade remains to be seen. State *de jure* ownership of the land could provide the foundation for recollectivisation of China's countryside in the future.

CONCLUSION

A cultural trait of the Chinese is that they are a historically minded people. Therefore it is understandable that during a period of transition China should look to its own past traditions as one of the sources for national restructuring. China is now also looking overseas, as it has done in the past. It remains to be seen how much of traditional China resurfaces and how much is transformed into a Westernised state. In any event, both aspects will be a part of the future, no matter what happens in the political arena.

FURTHER READING

Blunden, C. and Elvin, M. (1983) *Cultural Atlas of China*, Oxford: Phaidon.
   A good place for geographers with an interest in Chinese culture and history to begin their studies. This atlas is well-illustrated with colour photographs, and each map is accompanied by textual explanation.
Chi Ch'ao-ting (1963) *Key Economic Areas in Chinese History: as Revealed in the Development of Public Works for Water-Control*, New York: Paragon Books (reprint of 1936 edition).
   Although dated, this book still remains required reading for the student interested in China's historical geography. In the tradition of Karl Wittfogel, Chi sees water control as the main *raison d'être* of the Chinese state and attempts to focus this idea geographically. A reading of this book will give one a feel for China's regionalism and economic development history as well as political geography.
Elvin, M. (1973) *The Pattern of the Chinese Past: a Social and Economic Interpretation*, London: Methuen.
   One of the most significant Chinese history books of recent years. Elvin tries to anwer some basic questions about China: Why did the Chinese polity not disintegrate into smaller states? What led to China's economic revolution prior to the fourteenth century? Why subsequently did China fail to make further major economic progress? The book relies heavily on the work of Japanese economic historians and therefore provides a perspective on these problems not previously available in the English language.
Lattimore, O. (1988) *Inner Asian Frontiers of China*, Hong Kong: Oxford University Press (reprint of 1940 edition).

The classic work on China's northern frontiers. After a discussion of China's northern frontiers region by region, the second part of the book examines the role of frontier peoples in ancient China.

Whitney, J.B.R. (1970) *China: Area, Administration and Nation Building*, Chicago: University of Chicago Department of Geography Research Paper No. 123.

A work in political geography which has a strong historical–cultural focus. Whitney uses a geographical perspective to trace the evolution of China from an essentially cultural entity into both a cultural entity and a modern political entity in the twentieth century. The conclusion lacks the benefit of the post-Maoist experience but, none the less, the book provides good reading for the geographer.

Case Study 3.1

THE CHINESE LANGUAGE, GEOGRAPHY AND *PINYIN*

Terry Cannon

Chinese is written in a non-alphabetic way in the form of ideograms (often called 'characters'), each conveying in 'picture' form, a concept equivalent to an English word or part of a word. It is possible to represent Chinese in alphabetic form by using phonetic (i.e. 'sounds like') equivalents for the Chinese sounds. But it is not so straightforward: the sounds can be written variously by different people, depending on different interpretations of the best way to represent the Chinese sounds. In the past, the most common system of transliteration used in English-speaking countries was Wade-Giles, devised in the nineteenth century by those two people.

Making an acceptable method of putting Chinese sounds into Roman letters is made much more difficult by the tremendous variation in dialect and actual spoken languages in China. While the system of writing is the same across the country, people talk very differently even across relatively short distances. And, in effect, there are a number of major different spoken languages among the Han Chinese: a Shanghainese cannot normally be understood by people from Beijing, nor are Cantonese intelligible to those from the north (or again Shanghai).

In schools, a common spoken language is taught across the country, based on standard spoken Chinese (what we often call Mandarin, or Beijing dialect). It is important, too, in radio and television broadcasts. Obviously, such a set of language barriers has considerable significance for the spatial integration of this huge country, despite the common written form.

But the crucial problem is literacy. The ideogram system is a major barrier because at least 2,000 characters have to be learnt and understood for even basic reading materials to be any use. This, too, has a spatial component to the extent that teaching quality varies between areas, and especially between countryside and city. In the countryside, too, children tend to get less schooling, a problem which is now apparently worsening with the individualising of farming and need for labour from children.

For these reasons, a system using alphabetic equivalents of written Chinese has been of interest to the government for a long time. An official Romanized version of standard spoken Chinese has been developed and is used in school books and signs (with mixed success). It is called *pinyin*, and children learn this system in school. Many street signs and advertisements are produced using this Roman alphabet system as well. In theory, there is even a long-term intention to convert to this system and abandon the use of characters.

There is a further problem: the Chinese spoken languages are also 'tonal'. To convey meaning properly, each sound representing one of the ideograms should be pronounced with a particular tone, an inflection of the voice, on the vowel sound which forms part of each character when spoken. It is possible to indicate some tones by the use of 'accents' above the Roman letters. In standard Chinese there are four tones; in Cantonese seven or more, depending on the viewpoint of the linguist!

Chapter four

Environment: understanding and transforming the physical environment

Edward Derbyshire

The physical environments of China display a diversity which arises from a combination of the great size of the country (9.6 million sq. km), its position in middle latitudes on the eastern side of the Eurasian continent, its complex geological structure and its distinctive Pleistocene legacy of surficial sediments and soils. Many of the environments are harsh and, with easily tillable land making up less than 10 per cent of the total land surface, the burgeoning population needs to address a number of central issues of environmental science and to make some hard decisions affecting traditional practices.

TEMPERATURE, RAINFALL AND RUN-OFF PATTERNS

Climatic environments range from continental boreal in the northeast to moist tropical in the southeast, from mid-latitude desert in the north and west to alpine microthermal in the southwest. Much of the most densely populated parts of the country are dominated by the annual monsoonal cycle. Southeasterly moist tropical air-flows dominate the summer climate, and there is a consistent southeast to northwest decline in mean precipitation values, a pattern punctuated only by the higher mountain ranges of the north and west such as the Qinling Shan and Tian Shan (Figure 4.1a).

Climates

About one-third of China enjoys humid climates in which the naturally occurring vegetation is forest. This includes tropical rain forest and monsoon forests of the extreme south and southeast mainland and offshore islands, subtropical broad-leaved deciduous and evergreen (mixed) forest in northeast China from the eastern reach of the 'big bend' of the Huanghe to the Bohai Gulf, and cool-temperate *taiga* forest with mixed-temperate coniferous and broad-leaved forest east of Harbin in the extreme northeast. A number of forest–meadow associations occupy the subhumid belt making up about 14 per cent of China's land area. Examples include the

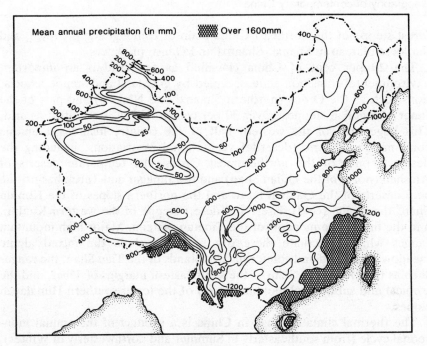

Figure 4.1a Mean annual precipitation

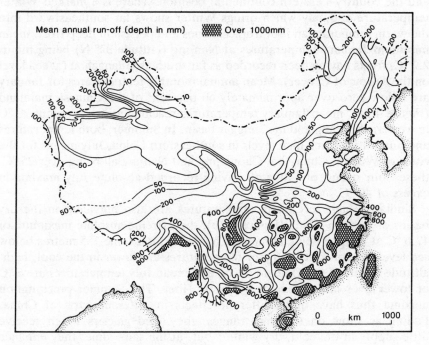

Figure 4.1b Annual surface run-off

forest steppe of the northeast China plain of Heilongjiang province, and the loess plateau in Shanxi, Shaanxi and Gansu provinces.

The 20 per cent of China classified as semi-arid is an important environmental transition zone occupied by grassland and steppe associations such as the Ordos (northern Shaanxi and Ningxia) and part of the Inner Mongolian steppes. Over 30 per cent of China is classified as arid. This huge area of the western half of the country includes a unique combination of xeric (desertic) vegetational associations, including the desert steppes of the northernmost Ordos, the temperate deserts of Alashan (north of the Qilian Shan) and the Junggar and Tarim basins, the cool deserts of the Qaidam basin and the northern slopes of the Kunlun Shan, the alpine (cold) desert and desert steppe of the western Kunlun, and the dry alpine meadow of the Tibetan plateau. Major high mountain ranges which punctuate these dry regions include the mixed alpine meadow, grassland and coniferous woodlands of the Tian Shan, the ranges such as the Hengduan Shan of the southeastern margins of Tibet, and the tropical and subtropical montane forest of the lower southern Himalayan slopes.

The thermal climatic range in China is a product of the annual monsoonal cycle (from southeasterly in Summer and northwesterly in Winter), the orography (with progressively higher mean altitudes towards the west), and the country's eastern continental position. There is a marked Winter temperature anomaly which brings Winter snows far southeastward into the warm temperate and subtropical zones and persistent deep frost, mean minimum January temperatures at Nanjing (latitude 32° N) being minus 2.2 °C, frosts having been recorded as far south as Guangzhou (at sea level on the Tropic of Cancer). Mean annual monthly temperatures for January are below zero over approximately 60 per cent of the Chinese mainland (Figure 4.2a), mean January temperatures reaching as low as minus 20 °C in western Xinjiang and the Junggar basin. In Summer, both temperatures and humidities reach high levels in southeastern China, July means for the cities of Wuhan, Changsha, Chongqing and Nanjing all being over 25 °C, these 'four ovens' of China all having recorded absolute July maxima in excess of 40 °C.

Similarly, high July mean temperatures are also prevalent in the dry basins of western China (Figure 4.2b), the extreme absolute maximum of 47.6 °C at Turpan being influenced by its location some 155 metres below sea level. In contrast, Summer temperatures are lower in the cool, high-altitude desert of the Qaidam basin, and mean July temperatures are 5 °C or lower over large areas of northern Tibet. The Summer precipitation maxima thus have quite different effects in different parts of China. Falling as snow on the high ranges, they feed glaciers which receive little supply in the cold, dry winters but, at the same time, they enhance the run-off into the headwaters of the major rivers and thus frequently

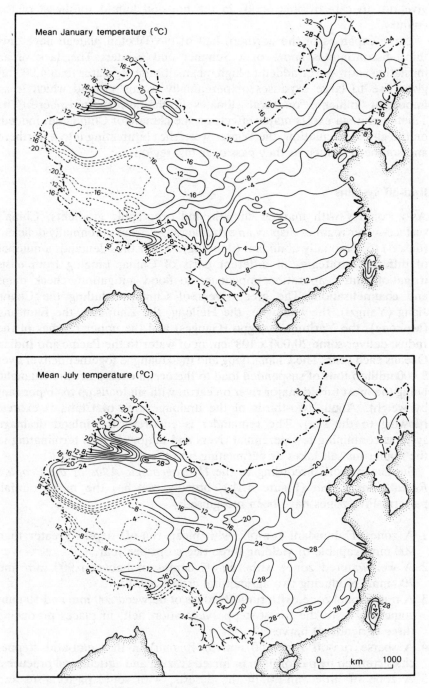

Figure 4.2 January and July mean temperatures

give rise to catastrophic floods in the hot and humid southeast of the country.

Large expanses of the northern half of the Tibetan plateau have only thin, discontinuous snow-cover Summer and Winter. This lack of an insulating snow-cover added to high mean altitudes (greater than 4,000 m) gives rise to large expanses of perennially frozen ground which is an important influence on local climates in the lower atmosphere. The Tibetan plateau exerts major effects on the climate of China and, indeed, on the global circulation, the Winter westerlies bifurcating into a northern and southern branch as they pass over this region.

Run-off systems

As a country with marked climatic and orographic gradients, China's surface-water regimes display an extreme range, from perennially deficient (desert) to seasonally abundant (flooding). Water thus demands a number of different strategies in different parts of China, ranging from oasis irrigation and evaporation impedence to flood mitigation, check dams and channelisation. The great rivers of China, including the Chang Jiang (Yangzi), the Huanghe, the Heilong, the Zhujiang, the Lantsang (Mekong), the Yarlung Tsangpo (Ganges) and the upper reaches of the Indus, deliver some $20,000 \times 10^{8^8}$ cu. m of water to the Pacific and Indian Oceans each year. The Chang Jiang and the Huanghe together deliver over 3,000 million tons of suspended load to the ocean each year, the Huanghe being the most turbid major river on earth, with silt loads up to 46 per cent by weight. About two-thirds of the drainage area of China is exoreic (flowing to the sea). The remainder is endoreic, the inland drainage systems containing few perennial rivers and major streams terminating in the numerous salt lakes or permeating into the desert surface.

Surface run-off, according to the *Hydrogeologic Atlas of the People's Republic of China* (Figure 4.1b), broadly matches the mean rainfall pattern. Five zones have been recognised:

1 A zone of abundant run-off, with mean run-off depths greater than 900 mm, capable of yielding three rice crops per year.
2 A well-watered zone, with run-off depths of between 900 mm and 200 mm, producing two agricultural crops each year.
3 A transitional zone, with run-off values of between 200 mm and 50 mm, coinciding with the forest-steppe vegetation belt, in places producing three agricultural harvests.
4 A sparse run-off zone (50 mm to 10 mm) in the semi-arid steppe, characterised by pastoralism or mixed grazing and agricultural practices.
5 A zone of little run-off in the deserts, with some pastoralism and agriculture sustained only by oasis waters.

In general, the seasonality of run-off produces a situation in which there is too much water in the Summer half year, and too little in Winter. The driest season with lowest surface run-off values is usually December to February: in central Gansu province at Lanzhou even the Yellow River turns blue, the suspended load being even more reduced than the run-off. There is a tendency for run-off to increase in the Spring throughout China, but there is a marked increase in central China as a result of Spring rains, and in northernmost northeast China and northern Xinjiang, where rapid snowmelt is responsible. Summer sees the heaviest run-off over most of China as a result of monsoonal precipitation, rapid and widespread snowmelt and ablation of glaciers, and the melting of frozen ground. In some areas, such as northern China, the Kunlun Shan and parts of Gansu province, Summer run-off may make up over 60 per cent of the annual total. Summer is the time of maximum flood hazard. Between 206 BC and 1949 there were 1,092 major floods in eastern China: these floods rise rapidly, flood waves travelling down the channel at velocities of more than 10 kilometres per hour.

MEGA GEOMORPHOLOGY

A major constraint on the human utilisation of land is the fact that some two-thirds of the country consists of high plateaux and mountains, no less than one-quarter of China's area lying above an altitude of 3,000 m. In the Chinese geological literature, the following four main altitudinal levels or 'great steps' are recognised (Figure 4.3):

1 The Qinghai–Tibet plateau, with an average altitude of 4,000 m.
2 A zone of plateaux and intermontane basins averaging between 1,000 m and 2,000 m lying between the Tibetan plateau in the west and the Da Hinggan, Taihang and Wu Shan mountains in the east.
3 Hill country and plains, generally less than 500 m above sea level and including the great plains of eastern China – the North China plain, the Northeast China plain, and the Chang Jiang plain.
4 The continental shelf of the Bohai Gulf, the Yellow Sea, and the East and South China Seas, with depths of less than 200 m.

However, the units of this classification contain considerable diversity arising from complex bedrock and structural characteristics, an active tectonic history including the present-day (neotectonics), and a distinctive geomorphic evolution during the last 2.5 million years.

Ancient crystalline rocks (granitic, metamorphic and metasedimentary) underlie many of the high mountain ranges including parts of the Kunlun, Qilian, Qinling and Tian Shan. There are extensive tracts of Mesozoic and Cenozoic continental rocks, mainly of fluvial and lacustrine origin, among which redbeds, mainly of sandstone and conglomerate, are prominent.

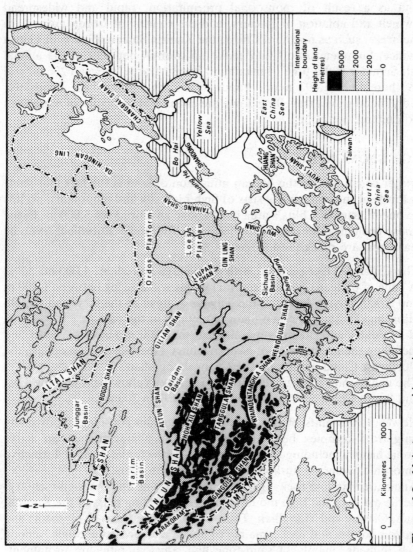

Figure 4.3 Main topographic regions

There are some formations of considerable economic importance including, for example, the coals of Jurassic age in Gansu province. Lime-rich rocks (limestones, dolomites, marbles) cover over one million sq. km, mainly in the south of the country. Extensive limestones of Palaeozoic age are found in a belt from Sichuan through Hunan and Hubei to Zhejiang provinces where there is a well-developed karst morphology, although this is not so evident on similar limestones to the (cooler) north in Shanxi, Hebei and Shandong. The most striking limestone landscapes, however, are in the very thick (over 8,000 m) limestones of Guangxi and the tower karst lands of Guizhou and eastern Yunnan developed on thick Permian and Carboniferous limestones and dolomites, the best-known area being around the city of Guilin. The karstlands south of the Chang Jiang are important for their well developed sub-surface drainage systems which often lead to local surface water shortages.

Structural trends

Two well-developed structural trends provide the framework of China's large-scale geomorphology. East of a line running through the Liupan and Hengduan mountains, the dominant structural trend is southwest–northeast, yielding a series of mountain blocks and inter-montane plateaux and basins – the Ordos platform, the Sichuan basin, the plateau of eastern Yunnan, the North and the Northeast China plains, the Shandong mountains, the Nanchang basin, the Huang Shan, the Wuxi mountains, the East China Sea and the mountains of Taiwan. West of the Liupan–Hengduan line, the dominant structural trends are east–west and northwest–southeast. The Tibetan plateau is bounded in the south by the Himalaya, culminating in Mount Qomolangma (Everest) at 8,848 m, and in the north by the Kunlun Shan. Its surface is surmounted by a number of mountain ranges rising above 6,000 m, such as the Nyainqentanglha, Tanggula and Gangdise mountains. North of the plateau is a series of broadly east–west trending block faulted ranges and basins – Altai mountains, Junggar basin, Tian Shan, Tarim basin, Altai Shan, Qilian Shan and Qaidam basin.

Seismic activity

In December 1988 two powerful earthquakes struck Yunnan in the southwest. This is an area of high seismic hazard, as much because of the instability of the mountain slopes in this region of high relief as because of the impact of the tremors themselves. Although the most powerful to affect the country for more than 10 years, there were fortunately relatively few deaths (the immediate figure quoted was around 800). However, nearly half a million homes were destroyed and farming livelihoods severely disrupted.

A number of clearly defined, seismically active belts can be traced across

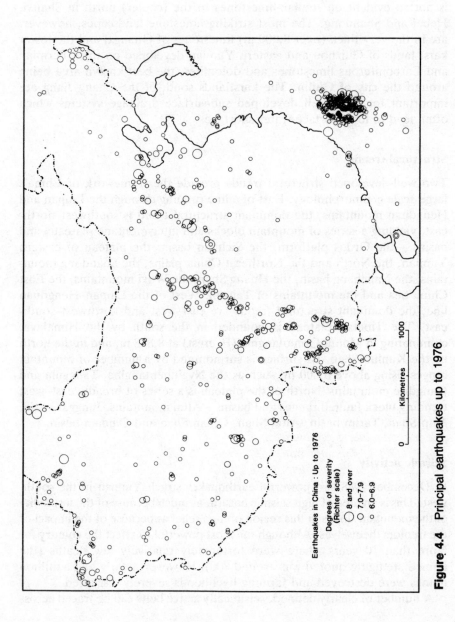

Figure 4.4 Principal earthquakes up to 1976

Earthquakes in China : Up to 1976

Degrees of severity
(Richter scale)

○ 8 and over
○ 7.0–7.9
○ 6.0–6.9

0   1000

Kilometres

China, there being an impressive record of earthquake incidence, location and damage over several centuries, a vivid testimony to the mobility of the basement of this country (Figure 4.4). Of recent earthquakes, the most destructive have occurred in the northeast around the Bohai Gulf. The worst, in 1976, killed nearly a quarter of a million people in and near the city of Tianjin and badly affected even Beijing.

Earthquake shock is a particular risk in areas mantled by thick, superficial sediments of Quaternary age, notably glacial deposits and wind-blown silts. It is not surprising that neotectonics and seismology are the concern of two major divisions of the Chinese Academy of Sciences, employing several thousand people, and that, previously, considerable effort was put into amateur monitoring systems.

THE QUATERNARY LEGACY

The Quaternary period, spanning a time from approximately 2.4 million years to the present, was remarkable for the environmental changes which have bestowed an important legacy upon China in the form of huge expanses of wind blown silt (called 'loess'), glacial deposits, huge alluvial plains, expanded deserts and, in Tibet, the largest expanse of perennially frozen ground outside the polar regions.

Glacial and periglacial areas

The geologically recent uplift of Himalaya and the Tibetan plateau has also resulted in the persistence of a glacial ice cover of some 56,471 sq. km, over 80 per cent of it lying within the borders of Tibet. Glacier distribution in China is strongly controlled by orographic factors, and most glaciers are of continental low-activity type, including most of those in the Altai, Tian Shan, Qilian Shan, Kunlun, Tanggula and Gangdise Shan. Maximum precipitation is in the warm season, over 70 per cent falling between June and August in parts of southern Tibet, much of which is rapidly lost from the glacier surface, enhancing the summer flood. Values of accumulation and ablation are low, however.

In contrast, the maritime monsoonal glaciers of southeastern Tibet, Sichuan and Yunnan have very high accumulation and ablation rates (c. 2,500 mm/yr and 10,000 mm/yr respectively), resulting in an active glaciological regime with rapid flow rates and high meltwater discharge values. In the dry west, altitudinally controlled morphoclimatic gradients are strong, from mountain ice-caps on the ranges, through a periglacial zone, then to the semi-desert lower mountain foot and finally to deserts of sand and gravel ('gobi'). Meltwaters from such glaciers and snowfields are a fundamental resource for the farmers and herdsmen around the margins of the Junggar, Taklamaken and Inner Mongolian deserts where, typically, they descend to alluvial fans across which have been built reticular channel systems to sustain some of the richest fruit and vegetable oases in Asia.

Outside the glaciated areas, perennially frozen ground extends over more than 2 million sq. km, approximately 22 per cent of the land area of China. Although most permafrost (1.5 million sq. km) is located in the Tibetan plateau, extensive occurrences have also been found in the Qilian, Altai and Tian Shan in the west and in the Da Hinggan Ling mountains in Heilongjiang province where it reaches 100 m in thickness and extends as far south as latitude 46° 36′ N. Across Tibet, the lower limit of continuous permafrost is 4,350 m in the Kunlun and 5,714 m in the Himalaya, a mean northwest–southeast gradient of 11 m per degree of latitude. Areas of permafrost are found as much as 1,300 m below the permanent snowline in the north and 800 m below it in the south. Much of the permafrost in Tibet is discontinuous south of 35° N, but landforms and structured soils are important throughout the plateau (patterned ground, blockfields, erected boulders, blockstreams, cryoplanation terraces, aufeis, pingos and a variety of thermokarst phenomena). Nivation and avalanching are important processes, especially in the south.

Resultant hazards

The presence of ice in the ground and in the form of glaciers presents a number of environmental hazards which are taxing the resources and ingenuity of the Chinese scientific establishment based in a number of centres, of which Beijing, Lanzhou and Chengdu are of particular importance. The combination of a summer precipitation maximum and the rapid summer warming of the glaciated mountains of central Asia is particularly important from this point of view because it gives rise to several environmentally hazardous processes, including the catastrophic debris flows known in China as 'mud-rock' flows. The matrix of clays and silts allows pore-water pressures to build up during the melting of snow and ice such that large boulders can be supported and transported on slopes as gentle as 3 per cent at velocities of up to 22 m/sec. Such flows occur widely in western China from the Tian Shan in the northwest to Yunnan and Sichuan in the south, the Tibetan flows including four main types.

Catastrophic earth and debris flows are an annual event in parts of Yunnan and Sichuan. With bulk densities of over 2,000 kg/cu. m and velocities of more than 15 m/sec., they exert dynamic forces of up to 200 tonnes/sq. m. They are a continuing threat to forests, agriculture and a number of man-made structures, including the Sichuan–Tibet and China–Nepal highways. Many of the deaths ascribed to the December 1988 earthquakes in Yunnan appear to have been a result of villages being struck by landslides dislodged by the tremors.

One of the most intensively studied areas of debris-flow threat is the

Jiang-Jia ravine in the drainage of the Xiao River (a tributary of the Chang Jiang) to the north of Kunming (Yunnan province). About 250 debris flows have been recorded here over the past twenty years, twenty-eight of these occurring in a single year. The total volume of debris involved has been estimated at between 30 and 35 million cu. m per year. The maximum debris-flow discharge (c. 2,800 cu. m/sec.) is more than five times greater than the fluvial discharge of the Xiao River. Such flows have been occurring for at least three centuries and may have increased as a result of deforestation for fuel in the local copper-smelting works. The partial reforestation of the past twenty years has not provided the same degree of stabilisation as the primary forest.

One other effect of such debris flows is disruption of human activities downstream at the outfall into larger rivers. In addition to direct damage to trunk roads and railways, debris flow enhances the aggradation of the trunk rivers and is threatening inundation of the already frequently damaged Kunming to Dungchuan railway. In addition, the Jiang-Jia debris has dammed the Xiao River to a height of 10 m at least seven times this century, forming 10 km-long lakes each time, with consequent damage to crops and roads. Damming of rivers by landslides is a major threat in this region. Triggered by the 1933 Deixi earthquake, three landslides dammed the Min River in northern Sichuan: the failure of the 255 m-high landslide dam produced a 60 m-high wave that killed at least 2,400 people. More recently, the Yalong River in central Sichuan was dammed by a 68 million cu m landslide in 1967 at Tanggudong. This 175 m-high dam was breached and a 50 m-high wave (discharge maximum c. 53,000 cu. m/sec.) swept down the valley.

The great earthquake of 16 December 1920 in Haiyuan county in Ningxia reached an intensity scale of 8.0 (Mercalli) over an area of 50,000 sq. km, killing 234,117 people. It created a surface rupture 215 km long marked by fault scarps and displaced the bed of the Hongwewe River in Jingyuan county by almost 1 m, as well as giving rise to numerous landslides. For the Tianshui earthquake of 1654 (Mercalli magnitude 8.0) and the 1718 Tongwei earthquake (Mercalli 7.5), both near Haiyuan, 396 seismically induced landslides have been recognised from air-photo studies. Following the 1950 earthquake (Mercalli 8.5) at Chayu in southern Tibet, gullying and debris flows were initiated in an area in which they were previously infrequent.

In China, destructive mass movements are triggered by glacial lake bursts, snow and glacier meltwater floods, melting of ground ice and avalanches, as well as monsoonal rain and earthquake shocks. Very steep ('hanging') glaciers, high mean annual precipitation and strong Summer ablation giving high meltwater discharges, and the presence of abundant, uncemented, superficial deposits of till, periglacial debris and bedrock loosened by frost-shattering and earthquakes, are conducive to widespread

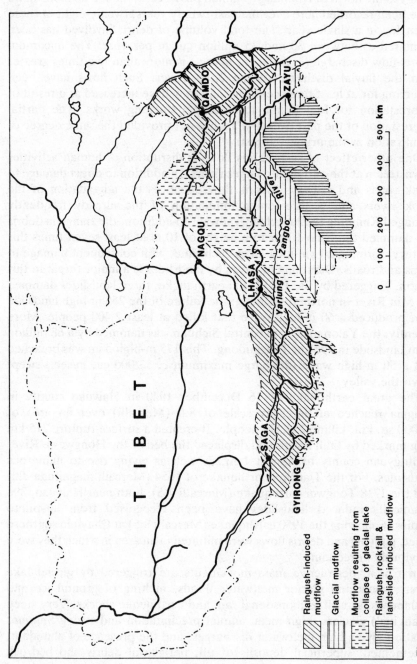

Figure 4.5 Unstable environments of southeast Tibet

Legend:

- Raingush-induced mudflow
- Glacial mudflow
- Mudflow resulting from collapse of glacial lake
- Raingush, rockfall & landslide-induced mudflow

QAMDO

ZAYU

NAGQU

LHASA

SAGA

GYIRONG

Yarlung Zangbo River

TIBET

0 100 200 300 400 500 km

glacier-related debris flows. On the south-facing slopes of the Mount Bogda region of the Tian Shan, for example, there is a close correlation between debris flow and glacier retreat.

Dumping by glacial mass flow processes of about 5 million cu. m of debris into a frontal glacier lake (the Damenlahai Lake) in Tibet in 1964 produced a total estimated flood discharge through the breached moraine of 3,698 million cu. m. The consequent mudflow destroyed a forest, buried twelve dwellings and submerged 2.2 km of the Sichuan–Tibet road. The Niyang River was blocked for ten hours and a 3 km-long lake created. The combination of frequent snow avalanches and a 'store' of unconsolidated glacial debris provides conditions for recurrent debris-flow problems in the Guxiang valley near Bomi (southeast Tibet). In 1964 a debris flow involving 3 million cu. m of debris occurred within a period of fifty-seven hours, obliterating crops, dwellings and stretches of highway. The debris and slope systems in monsoonal southeastern Tibet, Sichuan and Yunnan are amongst the earth's most dynamic geomorphic environments: much of the debris involved is of glacial and periglacial origin and all is of Quaternary age. (Figure 4.5).

The loess areas

The loess lands (Figure 4.6) provided the cradle for the rise of Chinese agriculture and civilisation and continue to provide a vital economic resource to the present day. They are estimated to cover 631,000 sq. km, or 6.6 per cent of the area of China, extending from Jilin province in the northeast through Liaoning, Hebei and Henan to the western peripheral belts around the Altun, Altai, Kunlun, Qilian and Tian Shan ranges, and including the thickest and most continuous deposits in the loess plateau of Shanxi, Shaanxi, Gansu and Ningxia in the middle reaches of the Huanghe. The loess reaches its greatest known thickness on earth (335 m) at Jiuzhoutai mountain near the city of Lanzhou. The loess plateau lies in a climatic transition zone between the humid monsoonal south and east and the deserts of the north and west. It is a region of net moisture deficit and, with altitudes of between 500 and 2,000 m and a continental interior location, winters are very dry and cold. The spring season is usually also dry, up to 70 per cent of the annual precipitation being concentrated in the months July, August and September.

Given the behavioural properties of loess and the geomorphology of loess terrain, these climatic conditions pose severe problems for farming and land conservation throughout this key agricultural region (see Case Study 4.1). In addition, the negative moisture balance over most of the loess lands means that irrigation has been essential to agriculture. In general, irrigation has been applied on a scale and on such surfaces (e.g. low terraces) as not to constitute a major erosional risk. More recently,

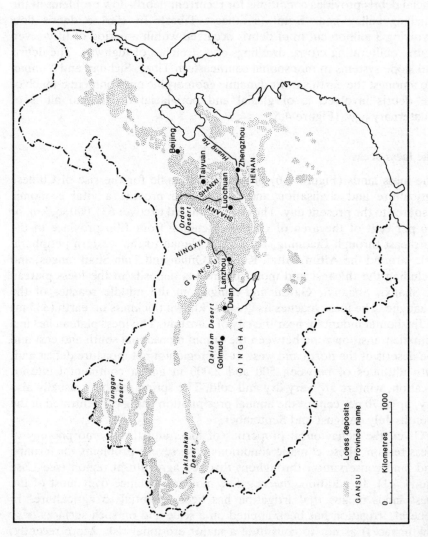

Figure 4.6 Loess deposits

however, the scale and extent of irrigation has increased, in some cases involving pumping schemes which deliver river water to high, hilly loess lands 200 or even 300 m above river level, as in eastern Gansu. There is little risk if such water is applied using spray methods, but many schemes have employed channel irrigation and this is beginning to cause problems which arise from local saturation followed by hydroconsolidation of the loess. Some high loess surfaces covered by extensive channel irrigation schemes, such as can be seen in eastern Gansu, have produced, in just a few years, measurable collapse of ancient buildings which had been stable since they were constructed many centuries ago.

On the desert margins of the loess country, underlain by sandy loess or sheet and dune sands, both the extent and intensity of gully erosion are much less, and landslides are rare. Sand encroachment in cultivated oases is a local problem in such areas as the Huanghe (Yellow River) valley in Ningxia, in Inner Mongolia, and at Dunhuang in western Gansu, but dust storms are a greater problem over large areas of northern and eastern China (Figure 4.7). There is evidence that both dust storms and sand

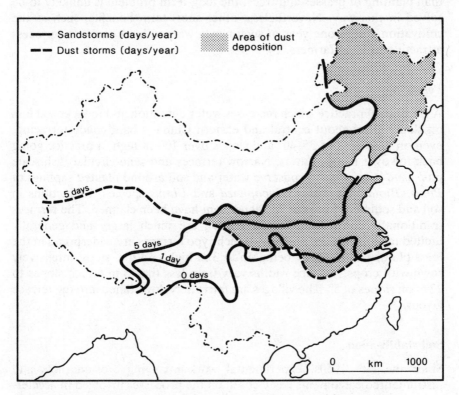

Figure 4.7 Sand-storms and dust-storms distribution and frequency

encroachment increased in historical times and that deforestation and the extension of crop cultivation into climatically marginal areas in the north of China have played a critical role in this process of desertification. It has been estimated by the Institute of Desert Research (Chinese Academy of Sciences) that 120,000 sq. km of desert expansion has occurred largely as a result of human activity in historical times.

REMEDIAL MEASURES

Measures designed to reduce run-off and mitigate debris flows over the past decade include both vegetational and engineering (structural) types. Masonry check dams, drainage diversion cuts and both stream-bed riffles and masonry and earth embankments are widely used to contain flood-water and to attenuate the effects of debris flows and smaller landslides. Former retrogressively eroding gullies in the loess plateau have been turned into sediment traps and over-all gradients reduced by artificially accelerating the hydroconsolidation process in a controlled series of operations. Although more effective in reducing soil loss in the short term than planting of grasses and trees, the long-term problem is unlikely to be solved in this way. Nevertheless, other short-term benefits, such as the cultivation within one year of such newly warped land, make the process attractive to local farmers.

Terracing

A traditional practice which increases water detention and reduces soil loss on slopes throughout central and eastern China is hand-made terracing, even on gradients of 35–40° and slopes over 100 m high, a practice going back for over five centuries. Narrow terraces and semi-circular 'fish-scale pits' have been used to conserve water and soil around planted saplings of *Pinus, Biota, Astragalus, Hypophaea* and *Ulmus* species. Reductions in soil and sediment losses of 75–90 per cent have been claimed. The terraces traditionally constructed for cultivation are much larger and generally limited to slopes of 25° and less. Bench-type terraces are widespread in the loess plateau, the 'risers' being up to 3 m high and the 'steps' intensively sown with crops. Terrace widths vary from less than 5 m on 25° slopes to 22 m on slopes of 5°. The villages are intricately dove-tailed into the terrace layout.

Soil stabilisation

Since the late 1950s, experimental work involving erosion plots and instrumented catchments has focused on the processes involved in acceler-ated gully erosion, infiltration, and sheetwash. This is especially so in the loess plateau, where much work has been done on the role of different plants in improving the stability of the soil. Forage crops significantly increase the organic content of loess soils at root level and so improve the content of water-stable aggregates. Alfalfa is particularly effective, a combination of perennial forage crops with alfalfa and *Dactylis glomerata* giving the highest yields and optimal aggregate percentages at all depths between 10 and 40 cm.

Rilling and gullying of loess soils are impeded by an appreciable moss cover but only to a depth of 5 cm. Moreover, moss-covered surfaces are likely to crack during dry periods and break up into readily rolled crusts. The most effective condition against rill and gully erosion is provided by the right combination of plants, especially by their rooting characteristics. Measures designed to minimise loss of moisture by evapotranspiration include the traditional one of covering all soil between each cultivated plant with river pebbles and, more recently, by covering crops with plastic sheeting. Plastic hot-houses also protect plants from hailstone damage and play a part in minimising soil detachment and erosion. In the northern half of the loess plateau, where soil loss by wind erosion is substantial, the planting of shelter belts has helped reduce deflation losses. Native species of *Hedysarum, Astragalus* (a fodder legume), *Caragana microphylla* (a legume with high fuel value which provides both good lumber and valuable fodder as well as having long tap-roots), black locust (*Robinia pseudo-acacia*), with its good fuel and lumber qualities, and *Larix* spp (larch), together with imported *Salix* spp (willow), have proved particularly successful. Looking to the future, there are plans to plant trees in suitable sites over a strip of land 7,000 km long across northern China as a composite human-made forest or 'green great wall'.

Flood control

Although the history of attempts to control the two great rivers of China goes back to the Han dynasty, the history of flood disasters continued until the last two or three decades. The Huanghe (Yellow River) is known to have changed its course at least twenty-six times in the past 4,000 years, and both the Huanghe and the Chang Jiang (Yangzi) River below the gorges pose recurrent flood threats by the shifting of meandering and braided channels. Since 1949, more than 2,000 km of dikes have been built along the lower Huanghe and flow velocities reduced by training-walls, river-bed baffles and piers. The basin of the Chang Jiang River receives about three-quarters of China's total mean annual precipitation. Although there is a summer maximum, the flood peaks in the tributary systems occur at different times, and large lakes in the middle and lower reaches, such as the Tungting Lake (with an area of more than 3,000 sq. km), provide some natural regulation of the flood peaks.

Although much of the middle and lower reaches of the river has a variously incised meandering channel, flooding was frequent prior to the hydraulic works of the past four decades: seven major floods occurred in the past century, that of 1931 killing 135,000 people. Principal remedial measures on the middle and lower Chang Jiang River have been bank-strengthening, the digging of flood-diversion channels and reservoir construction. Between 1949 and 1956 some 3,800 km of diking was built, involving 2.8 million cu. m of stonework.

A number of large reservoirs, including, for example, the Tangkiangkou on the Chang Jiang River, and the major works at Liujiaxia (with China's largest hydro-electric power plant) and Sanmenxia on the Huanghe River, have been constructed partly to provide some measure of flood control, as well as acting as sources of irrigation water and electrical power. There has been a considerable degree of success, although the long-term problem of reservoir filling by siltation remains, especially in such heavily laden systems as the Huanghe River. This is already shown by the curtailment of power generation at Sanmenxia. Problems of this and other kinds may be anticipated if the proposed controversial Three Gorges (Sanxia) reservoir on the Chang Jiang is pursued, with its 200 m-high, 2 km-wide dam impounding a lake some 500 km long. Of particular interest is the possibility of such a change in the hydrogeological conditions in the sedimentary rocks of the Chang Jiang slopes increasing the landslide hazard.

Problems of data inadequacy

However, taking China as a whole, solutions to problems raised by landslides, debris flows, floods, glacier surges and bursts, and permafrost melting must await the acquisition of better data. This should facilitate a better understanding of processes and critical thresholds, and be a basis for three-dimensional landscape modelling. Glacier inventory and research on the thermal regime, behavioural properties and areal extent of permafrost have been given a high priority in China, especially since 1976. Great stretches of the highway from Golmud to Lhasa overlie perennial frozen ground. Despite the recent tarmac-surfacing of this road, a satisfactory equation to describe the thermal stability of the substrate is still sought: it thus seems certain that recurrent problems will be encountered in its maintenance.

In the loess lands, given the dearth of financial resources and the lack of appropriate experience within some of the local scientific communities, there has been a tendency to resort to technological solutions on a local ad hoc basis. Although a beginning has been made, more research into the relationship between laboratory test characteristics, field performance (such as infiltration rates, etc.), bulk field engineering properties, and geomorphic form is needed. However, utilisation of a large proportion of the loess country demands techniques which are not easily mechanised. It is difficult to see how significant mitigation of loess-gullying, collapse, slide and flow can be achieved without substantial changes in age-old agricultural practices, together with significant increases in tree and perennial herb cover.

CONCLUSION

Improved management of the varied fluvial, loessic, desert, periglacial and glacial landscapes of China constitutes one of the world's greatest environmental challenges. The country has a cultivable area of only 10 per cent but a population of 1,000 million inhabitants. Only 5 per cent of them live in the half of China's territory made up of the plateaux, mountain ranges and deserts of the west. The degree to which this challenge is met will largely determine the chances of raising the economic well-being and political influence of one-fifth of the world's people.

ACKNOWLEDGEMENT

Some of the account of remedial measures rests heavily on an unpublished manuscript on 'The main types of water erosion and their related factors in the loess plateau' given to me at Wugong by its author, Professor Zhu Xianmo.

FURTHER READING

Biswas, A.K. *et al.* (eds) (1984) *Long Distance Water Transfer: a Chinese Case Study and International Experiences*, Riverton: T.Y. Cooly.
Greer, Charles (1979) *Water Management in the Yellow River Basin of China*, Austin University of Texas Press. Considers the problems posed by the flooding of the Huanghe (Yellow River) and analyses pre- and post-1944 attempts to manage the river.
Lee, J. S. (1939) *The Geology of China*, London: Murby. Until Yang *et al.*, (1986) the only major study in English of China's geology.
Liu Tungsheng *et al.* (1985) *Loess and the Environment*, Beijing: Ocean Press.
Pannell, C. W. and Salter C. L. (1985), *China Geographer No.12: Environment*, Boulder: Westview Press. A collection of papers by Chinese and American geographers, including problems of the loess region, agriculture, urban environmental quality, Beijing's water problem, fisheries, wildlife protection and others. There is also a chapter on how the earthquake risk has been managed, including information on precautionary planning and prediction work, for which China is famous.
Yang, Zunyi, Cheng Yuqi and Wang Hungzhen (1986) *The Geology of China*, Oxford: The Clarendon Press. A survey of Chinese geology by leading Chinese specialists covering stratigraphy, magmatic and metamorphic rock and geotechnic development.
Zhao, Songqiao (1986) *Physical Geography of China*, Beijing: Science Press with New York and Chichester: Wiley. Produced through the Chinese Academy of Sciences, this represents a pulling together of contemporary Chinese research.

Early chapters survey different aspects of climatology, geomorphology, surface and ground water. The second part consists of regional case studies. Over all, a very factual/descriptive approach.

Case Study 4.1

LOESS AND THE LOESS PLATEAU OF NORTH CHINA

Edward Derbyshire

Loess consists of generally unbedded, coarse-to-medium silts (grain size of dominant fraction in the range 0.01–0.06 mm). Because it is deposited by free fall from the air, it has low bulk densities (high porosities). Its mineralogy is rather simple, quartz being dominant with ancillary feldspars and micas. Calcium carbonate content, occurring as dispersed grains and as small concretions known as loess 'dolls', is low in the semi-arid northwest (7–16 per cent) but rises to 16–30 per cent in the more humid southeastern parts of the plateau in southern Shaanxi and Shanxi. The high porosities and low mean moisture contents give rise to high porewater tension and this, together with the presence of calcium carbonate occurring as coatings and cements between silt grains, results in the coarse columnar structure and vertical slope-faces characteristic of loess (which permit its excavation for use as 'cave houses' in much of the region).

The loess, deforested for centuries, has been widely reworked and redeposited by colluvial (slope) processes and by rivers (to form extensive stretches of loessic alluvium). The huge alluvial plains of northeastern China in the provinces of Henan, Hebei, Shandong, Anhui and Jiangsu are largely re-deposited loess, and the Huanghe (Yellow River) is named for its loess sediment load. The loess rests on red clayey silts, sands and gravels which contain a late Pliocene fauna, including the ancestor of the horse (*Hipparion* spp). Recent work on the palaeomagnetism of the classic loess plateau profile at Luochuan (Shaanxi) has shown that the base is about 2.4 million years old, suggesting that loess deposition coincided with the shift to cooler, drier conditions at the end of the Pliocene, a finding consistent with the results derived from deep ocean cores. The Chinese loess is therefore a phenomenon of the Quaternary.

Given the relatively unprotected devegetated surface of much of the loess region, owing either to an incomplete natural shrub cover (in the drier north and west) or to the intensive cropping patterns, loess is everywhere subject to erosion by water. Splash erosion by raindrop impact affects both dry and agriculturally moistened slopes, contributing to mechanical illuviation (washing in) and sheet erosion. The combination of rainbeat and overland flow (predominantly of the surface saturation type) breaks up the weak soil aggregates and both dissolved and suspended solids are transported downslope. This tends to render the surface layer more sandy and therefore less valuable agriculturally.

Gully development and extension is very active on the loess plateau, sediment concentrations in gullies within the Malan loess reaching 700–1,000 kg/cu. m (48.7–61.5 per cent by weight) during Summer storms (Gong and Jiang 1979). Loess erosion values of 18,600 t/sq. km/yr have been recorded in parts of Shaanxi. Raindrop impact, rilling, gullying, slurry flow and sliding of loess blocks downslope combine to produce very high mean values of loess erosion, e.g. in the Dali River in Shaanxi in the decade 1959–69, the mean annual value was 19,600 t/sq. km. It has been calculated that the mean loess thickness eroded from the loess plateau in one year is equivalent to the loess accumulated over a period of 100–300 years: the resource is thus diminishing rapidly, given present land-use practices.

A major and recurring problem arising from the tendency for loess to collapse instantaneously when saturated is the large-scale failure of the flowslide type and associated mass flowage akin to mudflows. Wet slides and loess flows tend to concentrate in the Summer wet season between June and August, events in August making up some 70 per cent of all such events in eastern Gansu. Almost one-third of the area of eastern Gansu province is subject to flowslides and mass flowage in thick loess, the largest failures being attributable to earthquake shock.

This is a region in which the rural population is concentrated in villages on valley floors beneath steep valley sides. The effects of large-scale failures in the loess are thus frequently catastrophic, destroying reservoirs, initiating soil erosion by gullying and burying both the cultivated fields and the villages themselves. Over 1,000 large landslides occurred in the highlands of eastern Gansu between 1965 and 1979, killing over 2,000 people. The 1983 Sala Shan landslide in southern Gansu was triggered by an earthquake. In less than one minute the slide buried four villages and 200 ha of farmland, a reservoir was destroyed, the river was pushed 0.5 km across the valley and 227 people were killed. In the Tian Shui prefecture in southeastern Gansu between 1964 and 1978 seven large debris flows destroyed 17,544 houses, covered 22,500 ha of cultivated land and killed 1,142 people.

REFERENCES

Gong, S. and Jiang, D. (1979) 'Soil erosion and its control in small watersheds of loess plateau', *Scientia Sinica* 22: 1302–13.

Chapter five

People: demographic patterns and policies

John Jowett

With more than 1,080 million people, China accounts for about one-fifth of the world's population. Its government considers that, in relation to available resources, the population is too large and growing too rapidly. A case of too many fertile people on too little fertile land. In attempting to understand concern over the population issue it should be borne in mind that on about 7 per cent of the world's arable land China now supports a population which, some 150 years ago, was supported by the entire globe. It is edifying to note that in the five years 1981–6 the increase in population (60 million) exceeded the total population of Britain. Since 1949 China's population has increased by over 500 million, an increase which exceeds the combined populations of the USA and the USSR. In the decade 1963–73 the population increased by 200 million. To meet the rising expectations of a population which has doubled since 1949 has presented a major challenge to the government of the People's Republic. In an endeavour to contain the threat posed by a large and rapidly expanding population, the government introduced its family-planning programme in the 1970s and its one-child policy in the 1980s.

SPATIAL PATTERNS

China's population is characterised by very uneven distribution, a predominantly rural location, and the numerical dominance of the Han nationality, despite immense territories in which other ethnic groups are significant (Figure 5.1). In terms of the environmental constraints operating within China it is not surprising that the population is very unevenly distributed. Provincial densities vary from less than two people per sq. km in Tibet to over 600 in Jiangsu province and almost 2,000 in the municipality of Shanghai.

The east–west divide

So uneven is the distribution that the border provinces of Heilongjiang, Inner Mongolia, Ningxia, Gansu, Qinghai, Xinjiang, Tibet and Yunnan

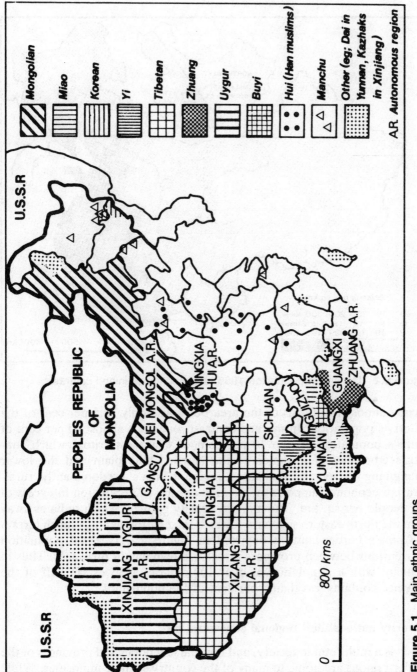

Figure 5.1 Main ethnic groups
Source: After Pannell and Ma (1983)

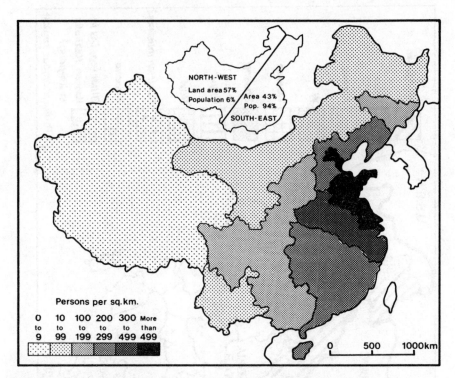

Figure 5.2 Density of population 1985, showing coast–inland contrasts

occupy almost two-thirds of the area but house only 12.5 per cent of the country's population. In marked contrast, one finds almost 40 per cent of China's population on the 10 per cent of Chinese territory which constitutes the alluvial lowlands of the North China plain and the lower Changjiang valley. In the latter region China's predominantly rural-agrarian economy supports an average density of population in excess of 400 people per sq. km. The marked density gradient which radiates away from the north eastern seaboard towards the environmentally harsh terrain of China's border-lands is depicted in Figure 5.2. Unequal population distributions between provinces are also shown within provinces. Thus in Sichuan, with a population of over 100 million, the western half of the province contains less than 5 per cent of the population.

Minority nationalities' regional concentration

China is a multi-ethnic society, and the sparsely populated provinces of the border regions are home to many of the country's ethnic minorities. While the Han account for over 93 per cent, there were, in 1982, fifty-five

Table 5.1 Demographic data for China and selected provinces, 1981–2 (Part 1)

	Total population[a] (millions)	Population increase 1964–82 (per cent)	Population density (people per sq. km)	Rural population (per cent)	Urban population (per cent)	Ethnic composition (per cent Han)
Beijing	9.23	21.5	549	35.3	64.7	96.5
Shanxi	25.29	40.4	162	79.0	21.0	99.7
I. Mongolia	19.27	56.3	16	71.2	28.8	84.5
Liaoning	35.72	32.6	245	57.6	42.4	91.9
Heilongjiang	32.6	62.3	69	59.5	40.1	95.1
Shanghai	11.86	9.6	1913	41.2	58.8	99.6
Jiangsu	60.52	35.9	590	84.2	15.8	99.8
Anhui	49.67	59.0	356	85.8	14.2	99.5
Fujian	25.87	54.4	213	78.8	21.2	99.0
Shandong	74.42	34.1	486	80.9	19.1	99.5
Hubei	47.81	41.8	255	82.7	17.3	96.3
Guangdong	59.30	46.6	280	81.4	18.6	98.2
Sichuan	99.71	46.6	176	85.7	14.3	96.3
Guizhou	28.5	66.5	162	81.1	18.9	74.0
Yunnan	32.55	59.2	83	87.1	12.9	68.3
Gansu	19.57	54.9	43	84.7	15.3	92.1
Qinghai	3.90	81.6	5	79.5	20.5	60.6
Xinjiang	13.08	79.9	8	71.6	28.4	40.4
CHINA	1,008.18[b]	45.2	105	79.5	20.5	93.3

Source: The 1982 Population Census of China.
Notes: [a] Year-end population totals.
[b] Includes 4.24 million servicemen who are excluded from the provincial totals.

officially recognised national minorities with a combined population of 67.2 million. Benefiting from improved living conditions and hence lower mortality, and spared from the rigours of the country's population-control policies, the ethnic minorities have increased rapidly. In the inter-censal period 1964–82 the population of the Han rose by 43.8 per cent, while that of the minorities increased by 68.4 per cent. This is one aspect of the marked regional variations in the pattern of growth (Table 5.1).

Coast–inland differences

Lower-than-average growth rates are characteristic of the urban munici-palities (Shanghai, Beijing, Tianjin), the more urban and more econ-omically developed provinces (e.g. Liaoning) and the densely populated coastal areas (such as Jiangsu and Shandong). Together these provinces make up that continuous arc of territory around the northeastern sea-board, from Liaoning in the north to Zhejiang in the south (Figure 5.2). On the other hand, the border provinces, mostly with much sparser populations, are characterised by high rates of population growth (Figure 5.3).

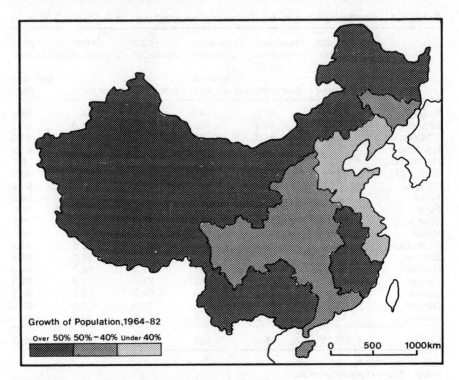

Growth of Population,1964–82

Over 50% 50% – 40% Under 40%

0   500   1000km

Figure 5.3   Growth of total population 1964–82

In these peripheral regions the boost in population numbers has come from a tolerance of higher rates of natural increase among the ethnic minorities coupled with a significant in-migration of (mainly Han) people. During the thirty years, 1949–79, the net migration into China's major receiving provinces is estimated at 25–30 million, with net movements of around 8 million into Heilongjiang, 5 million into Inner Mongolia, 3 million into Xinjiang, and major migrations into Qinghai and Ningxia (Case Study 5.1). The densely populated eastern provinces have provided the source areas for most of the migrants. Thus the net out-migration from Shandong province since 1949 is estimated to be 4 million, and in Shanghai the migration flows in and out of the city proper since 1949 have resulted in a net loss of 1.04 million.

Three contiguous eastern provinces – Anhui, Jiangxi and Fujian – combined relatively high densities of population with above-average rates of population growth (Tables 5.1 and 5.2). In these cases the population boost has come from above-average rates of natural increase (arising from below-average levels of social and economic development) and the overspill of population from those adjoining, most densely populated provinces of Henan, Shandong, Jiangsu, Shanghai and Zhejiang.

Table 5.2 Demographic data for China and selected provinces, 1981–2 (Part 2)

	CBR^a per 1,000	CDR^a per 1,000	NI^a per 1,000	TFR^b	Life expectancy	Per cent popn over 15 years	Median^c age
Beijing	17.55	5.78	11.77	1.589	72.06	22.38	27.20
Shanxi	20.31	6.54	13.77	2.385	67.87	33.37	22.97
I. Mongolia	23.11	5.77	17.34	2.621	67.01	35.52	21.11
Liaoning	18.53	5.32	13.21	1.773	70.90	28.71	24.60
Heilongjiang	19.79	4.95	14.84	2.062	68.40	34.89	21.54
Shanghai	16.14	6.44	9.70	1.316	73.13	18.16	29.23
Jiangsu	18.47	6.10	12.37	2.076	69.81	28.98	25.53
Anhui	18.73	5.20	13.53	2.799	69.54	36.15	20.17
Fujian	22.07	5.87	16.20	2.717	68.71	36.50	20.68
Shandong	18.84	6.26	12.58	2.104	70.25	31.04	24.55
Hubei	20.17	7.33	12.84	2.445	65.84	32.72	23.05
Guangdong	24.99	5.54	19.45	3.283	71.42	33.91	22.53
Sichuan	17.96	7.02	10.94	2.434	64.57	34.38	23.42
Guizhou	27.89	8.48	19.41	4.355	62.01	40.88	18.76
Yunnan	25.36	8.60	16.76	3.814	61.45	39.17	19.40
Gansu	20.12	5.72	14.40	2.728	66.09	36.32	20.10
Qinghai	26.65	7.48	19.17	3.927	61.43	40.56	18.54
Xinjiang	29.08	8.41	20.67	3.883	61.48	39.56	19.54
CHINA	20.91	6.36	14.55	2.584	67.88	33.59	22.9

Sources: The 1982 Population Census of China; Beijing Review 12.3.84. (total fertility rate); Guangming Ribao 24.8.86. (life-expectancy at birth); Statistical Yearbook of China 1986 (median age).
Notes: ^a CBR = crude birth rate; CDR = crude death rate; NI = natural increase.
^b TRF = total fertility rate (roughly equivalent to the average number of children born to each women in the age-group 15–49).
^c The median age divides the population in half, 50 per cent of the population are older than the median age, 50 per cent are younger.

Significant in-migration of military and other personnel into Fujian has occurred in connection with the disputed sovereignty of Taiwan.

Fertility variations

Patterns of internal migration provide only part of the explanation for the spatial variations in the provincial pattern of population growth. The western provinces are characterised by much higher levels of fertility than prevail in the more developed provinces of eastern China. Data on birth rates, total fertility rates and the natural growth rate of population are included in Table 5.2.

After fifteen to twenty years of strict population control policies aimed mainly at the Han, many of the border provinces (Guangxi, Tibet, Xinjiang, Ningxia) still record natural growth rates of population ranging from 2.0 per cent to 2.4 per cent each year. These are double those in most of the coastal provinces from Zhejiang northwards to Liaoning. Bearing in mind that a population growing at 2 per cent a year doubles itself in thirty-five years in contrast to the seventy years required by a population growing

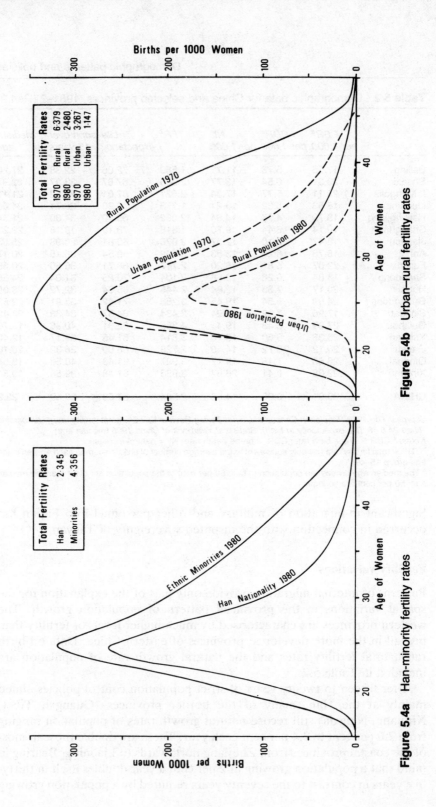

Figure 5.4b Urban-rural fertility rates

Total Fertility Rates	
1970 Rural	6·379
1980 Rural	2·480
1970 Urban	3·267
1980 Urban	1·147

Figure 5.4a Han-minority fertility rates

Total Fertility Rates	
Han	2·345
Minorities	4·356

at 1 per cent, it will be apparent that differential rates of natural increase offer an important explanation for the spatial variations in the pattern of population growth in China.

Lower levels of socio-economic development and the concentration of the ethnic minorities in the west explain the higher level of fertility in that region. Early marriage, early childbirth, a long childbearing lifespan and many births are characteristic of most of the ethnic minorities. A more reliable way of showing this than crude birth rates is to use total fertility rate (TFR). This is roughly equivalent to the average number of children born to women in the age-group 15–49. In 1981 the TFR for rural women of minority nationality was 5.049 compared with 2.758 for Han women, and in rural China nearly 30 per cent of the births to minority women were delivered into families which already had five or more children. As shown in Figure 5.4a, the greatest differences in fertility behaviour between the Han and the ethnic minorities is to found among older women. After the first child most Han women practise contraception, while the minorities continue to have children.

Development and demography

While the more relaxed approach towards family planning among the ethnic minorities offers some explanation for the high levels of fertility in the western region, it is also the case that in China, as elsewhere in the world, there is a strong negative correlation between levels of socio-economic development and levels of fertility. The lower levels of literacy, life-expectancy, urbanisation, industrialisation and income which prevail in the west are associated with higher levels of fertility. Conversely, in the coastal provinces around the northeastern seaboard, higher levels of social and economic development correlate with lower levels of fertility. Table 5.3 provides data for Liaoning and Guizhou to highlight the relationship between demographic behaviour and levels of socio-economic development in a coastal and interior province.

High levels of natural fertility combined with age-selective migration have generated a series of very youthful populations in the west. Many of the border provinces have around 40 per cent of their population under the age of fifteen (Guizhou, Yunnan, Ningxia, Qinghai and Xinjiang). In marked contrast, the demographically more mature provinces on the northeastern seaboard (Liaoning, Jiangsu, Zhejiang) have a juvenile population of under 30 per cent and in Shanghai the under-15s account for less than 20 per cent of the municipality's population (Table 5.2).

Table 5.3 Social, economic and demographic data for Liaoning and Guizhou

	units	year	Liaoning	Guizhou
Female literacy	%	1982	76.6	32.6
Life expectancy	years	1981	70.9	62.0
Infant mortality	per 1,000 births	1982	19.7	62.9
Ethnic composition	% Han	1982	91.9	74.0
Per capita GNP	Rmb	1982	966.0	217.0
Per capita GVIO[a]	Rmb	1981	1,276.0	156.0
Per capita GVAO[b]	Rmb	1981	238.0	145.0
Consumption level	Rmb/person	1984	488.0	258.0
Peasants' net income	Rmb/year	1983	452.0	225.0
Bicycle ownership	per 100 popn	1980	13.1	0.7
Radio ownership	per 100 popn	1980	12.0	0.7
Total population	millions	1982	35.72	28.55
Urban population	%	1982	42.4	18.9
Population increase	%	1964–82	32.6	66.5
Birth rate	per 1,000 popn	1981	18.5	27.9
Total fertility rate[c]		1981	1.8	4.4
Population < 15 years	%	1982	28.7	40.9

Sources: Statistical Yearbooks of China 1981–6; The 1982 Population Census of China.
Notes: [a] GVIO = gross value of industrial output.
[b] GVAO = gross value of agricultural output.
[c] TFR is roughly equivalent to the average number of children born to each woman in the age-group 15–49.

TEMPORAL PATTERNS

The failure to develop a coherent policy on population control in the 1950s and 1960s meant that throughout those decades it fluctuated between a pro- and an anti-natalist line. In the main, Mao's pro-natalist views held sway: 'Of all things in the world people are the most precious. Under the leadership of the Communist Party, as long as there are people, every kind of miracle can be performed. .... All pessimistic views are utterly groundless' (Mao 1969). In pursuit of such beliefs, sterilisation was prohibited and a strict control was placed on abortion.

The early programmes for population control in 1956–7 and 1962–6 were short-lived, had limited financial and political support and had very little impact on reducing fertility. It was not until the 1970s that the consensus switched from an optimistic view of people as producers to the more pessimistic view of people as consumers. At that point the leadership finally opted for an intensive programme of birth control.

Graphs of birth rates and death rates since 1949 highlight the fluctuating fortunes of China's population (Figure 5.5). In the period 1949–57 the major thrust of population policy was aimed at reducing mortality. The return of internal peace, the improvements in food production, distribution and consumption, the improvements in water supply and sanitation, and the substantial improvements in medical care combined to lower significantly the death rate in general and infant mortality in particular.

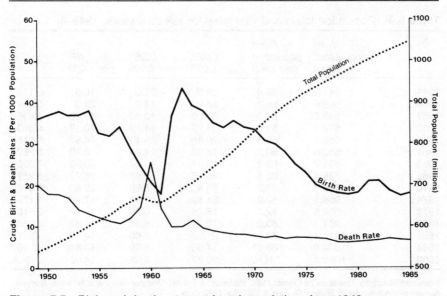

Figure 5.5 Birth and death rates and total population since 1949

Famine disaster

In the late 1950s and early 1960s the country experienced a major demographic disaster, one of the most devastating in the history of the world. A combination of man-made and natural disasters caused a massive decline in foodgrain production which triggered off severe famine conditions throughout China.

The demographic impact of the famine was such that by 1960 the soaring death rate surpassed the plummeting birth rate, and China's population went into decline (Figure 5.5). Over the two years 1960–1 China's official view is that the population declined by 13.5 million. Annual deaths probably rose from 11.5 million in 1957 to around 29 million in 1960 and over the four years 1958–61 China suffered some 25–30 million more deaths than might have been expected under normal conditions.

While everyone is aware that famine increases the death rate, it is less widely known that famines are also associated with major reductions in the birth rate. A combination of biological, psychological and behavioural mechanisms explain the dramatic decline of fertility at times of famine. In China's case famine conditions were responsible for halving fertility between 1957 and 1961, with the total fertility rate declining from 6.405 to 3.287 (Table 5.4). The annual number of births fell probably from over 27 million in 1957 to about 14.5 million in 1961. In total, some 30–35 million births were lost or postponed as a result of the famine.

Previous experience indicates that periods of famine-induced infertility

Table 5.4 Population totals and vital rates for selected years, 1949–86

	Total population[a] (millions)	Rural population (per cent)	CBR[b] per 1,000	CDR[b] per 1,000	NI[b] per 1,000	TFR[c]
1949	541.7	89.4	36.0	20.0	16.0	6.138
1953	588.0	86.7	37.0	14.0	23.0	6.049
1957	646.5	84.6	34.03	10.80	23.23	6.405
1959	672.1	81.6	24.78	14.59	10.19	4.303
1960	662.1	80.3	20.86	25.43	−4.57	4.015
1961	658.6	80.7	18.02	14.24	3.78	3.287
1963	691.7	83.2	43.37	10.04	33.33	7.502
1966	745.4	82.1	35.05	8.83	26.22	6.259
1970	829.9	82.6	33.43	7.60	25.83	5.812
1974	908.6	82.8	24.82	7.34	17.48	4.170
1978	962.6	82.1	18.25	6.25	12.00	2.716
1980	987.1	80.6	18.21	6.34	11.87	2.238
1982	1015.4	79.2	21.09	6.60	14.49	(2.71)[d]
1984	1034.8	68.1[e]	17.50	6.69	10.81	(2.16)[d]
1986	1060.1	N.A.	20.77	6.69	14.08	N.A.

Sources: Statistical Yearbook of China. 1986; Banister, J. (1984) One-per-Thousand National Sample Survey of Fertility, 1982; Beijing Review 2.3.87 (for 1986 data).
Notes: [a] Year-end population totals.
[b] CBR = crude birth rate; CDR = crude death rate; NI = natural increase.
[c] TFR = total fertility rate (roughly equivalent to the average number of children born to each woman in the age-group 15–49).
[d] Data supplied by Banister (1984).
[e] The greatly reduced rural population in 1984 is accounted for by a change in the rural/urban definition.

are often followed by a phase of hyper-fertility. This was certainly the case in China, for in 1963 the TFR soared to 7.502 and the crude birth rate probably reached fifty per thousand. Some 33.5 million babies were born in 1963, and in the three years from mid-1962 to mid-1965, over 90 million births were recorded.

The rapidly fluctuating number of births in the period 1957–65, trimmed by variable rates of survival, appear as series of peaks and troughs on the 1982 age structure of population (Figure 5.6). Such is the variation in the size of the cohorts that in 1982 there were 19.5 million 24-year-olds and 27.4 million 19-year-olds but only 10.7 million 21-year-olds. The dearth of population in the 20–23 age-group, survivors from the births of 1959–62, is a clear indication of the extent to which the famine impinged on Chinese levels of fertility and mortality.

Birth-boom effect

Past demographic problems are connected to present-day population policies. The baby-boom of 1963 reached the age of marriage and childbearing in the mid-1980s, and the growing awareness of the existence of this large, inbuilt momentum for population growth was one of the factors that led the Chinese government to initiate its one-child policy.

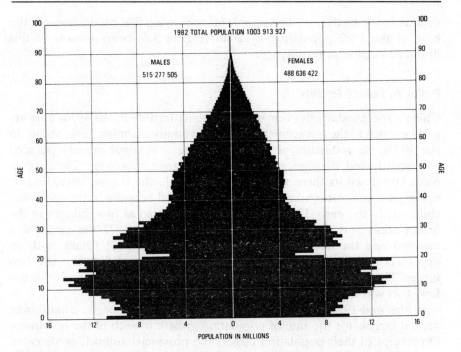

1982 TOTAL POPULATION 1003 913 927

MALES
515 277 505

FEMALES
488 636 422

POPULATION IN MILLIONS

Figure 5.6 Age structure of the population in 1982 (population pyramid)

Those born of the baby-boom in the 1960s are currently being required to severely limit their fertility in an attempt to restrain the rate of population growth and to contain the size of the country's population within the 1,200 million targeted for the year 2000.

In the aftermath of the baby-boom of 1963 the government began to seriously question the value of a rapidly growing population and sanctioned the introduction of family-planning programmes, especially in the larger urban centres. The outcome of these activities was a substantial reduction in urban fertility with the TFR down from its peak of 6.2 in 1963 to 3.1 in 1966. In rural areas the impact was minimal. This particular phase of the birth-control programme was brought to an end by the Cultural Revolution (1966–9).

THE POPULATION CONTROL PROGRAMME OF THE 1970s

Following the turmoil of the early phase of the Cultural Revolution, the new economic and social policies for the 1970s included a dramatically successful birth-control programme. So successful that there were 10 million fewer births in 1979 than in 1971 and the natural growth rate of population declined from 25.83 per thousand in 1970 to 11.61 in 1979. The information in Table 5.4 and the graphs in Figures 5.4 and 5.5 detail the decline in the birth rate. The significant narrowing that occurs towards the base of the 1982 population pyramid (Figure 5.6) bears witness to this dramatic decline in fertility.

Policy to reduce fertility

China's spectacular achievement in reducing fertility in the 1970s appears to owe much to the government-sponsored family-planning programme. In the 1970s the reduction of fertility became a national priority pursued consistently and vigorously through a wide range of policies. The government laid down its three reproductive norms in the slogan 'later, longer, fewer', meaning later marriage, a longer interval between births and fewer children. In the early-1970s 'fewer' was interpreted as two children in the urban areas and three in the countryside, but in 1977 the target was lowered and the norm was reduced to two throughout China, with an expectation that there would be a four-year interval between the two births. The new objective was encapsulated in the slogan 'One is not too few, two will do and three are too many for you.'

In the mid-1970s the Family Planning Commission began establishing annual targets for the rate of population growth in each of the provinces. On receipt of their population targets the provincial authorities allocated birth quotas to the counties and the process passed down the administrative chain until the targets finally reached the production team (an average of thirty families) or its urban equivalent. At this level, group meetings determined which women would be eligible to have a baby in the coming year and thus there arose the practice of 'giving birth in turn'. The authorised couples were issued with a planned-birth certificate and this had to be presented at all pre-natal examinations. The remaining married couples were expected to practise contraception and undergo abortion in the case of unplanned pregnancies. Contraceptives and family-planning operations were free and widely available to married couples, though there were indications of a rapid rise in abortions among unmarried women.

Contraception campaigns

Family-planning information is disseminated through posters, newspapers, radio, and especially through small group meetings where strong peer-group pressure is exerted. The major thrust of the educational programme has been aimed at changing traditional attitudes from a belief in early marriage and many children, especially sons, to an acceptance of late marriage and few children. Some indication of the extent to which traditional attitudes towards child-bearing were remoulded can be gauged from the fact that between 1971 and 1979 some 210 million birth-control operations (IUD insertions, sterilisations and abortions) were carried out.

The family-planning campaign of the 1970s raised contraceptive use to the levels presently experienced in the developed world, with almost 70 per cent of married couples of child-bearing age currently practising some form of modern contraception. Over 50 per cent of couples use IUDs, 35 per cent are protected by sterilisaton and about 8 per cent use the pill. Despite the high level of contraceptive use, induced abortion is extensively used.

Education and reduced family size

The government-sponsored family planning campaign was just one of the reasons for the success of the birth-control programme in the 1970s. Improvements in education, health, income and the changing role of women in society combined to make a very positive contribution towards changing attitudes to child-bearing. More education for women, for example, is one of the factors which generates a reduction in fertility. Educated women not only delay marriage but once married are more likely to understand and practise effective methods of birth control. In 1981, illiterate mothers averaged 4.74 births, those with a primary education 3.81, mothers with senior secondary education averaged 2.41 and those with a tertiary education had only 1.94 births. The substantial improvement in female education since 1949 undoubtedly made an important contribution to the decline of fertility in the 1970s.

Improved education increases the age of marriage and this has an important influence on levels of fertility. Early marriage is associated with high fertility and late marriage with low fertility. Later marriage certainly played an important role in lowering the birth rate in the 1970s. Calculations suggest that about a quarter of the decline in rural fertility in the 1970s is accounted for by the increase in the age of marriage. The average age of females at first marriage rose from 18.5 years in 1949 to about 23 years in 1979, with a particularly rapid increase occurring in the 1970s. The percentage of women who were married by the age of 20 declined from 70.9 per cent in 1965 to 24.9 per cent in 1980.

Age of marriage

There are substantial regional variations in the average age of first marriages, ranging from 27.8 and 25.8 years for men and women respectively in Shanghai to 23.9 and 21.4 in Qinghai. Throughout the period 1950–80 the minimum ages for marriage were as legislated in the Marriage Law of 1950, 18 for women and 20 for men. The increasing age of marriage over the past 30 years has been a response not to legislated change but to campaigns, social pressure, and the rapidly changing social environment. Given the widespread success in raising the age of marriage and the positive impact it had on reducing the birth rate it came as a major surprise when the new Marriage Law of 1980 raised the minimum age of marriage by only two years, 20 for women and 22 for men. In the early 1980s it therefore became legally possible for women to marry some three to five years earlier than had been administratively permissable in the late-1970s. The new Marriage Law, with its implied relaxation on the timing of marriage, generated a decline in the age of marriage and a temporary surge in the number of marriages. The considerable increase in the number of newly-weds produced the minor baby-boom of 1981–2 which is imprinted on Figure 5.6 and apparent from the data in Table 5.4. Given the vagaries of government policy, few newly married couples delay child-bearing. The rate of natural increase which had descended to 11.6 and 11.9 per thousand in 1979–80 climbed to 14.6 and 14.5 in 1981–2.

Provincial variations

On the other hand, analysis of provincial data shows a clear and negative correlation between levels of socio-economic development and levels of fertility. Provinces with higher levels of socio-economic development display lower levels of fertility and vice versa. Futhermore, the more developed urban areas show significantly lower levels of fertility than the less developed rural areas. Thus it should not be presumed that the spectacular decline of fertility in the 1970s resulted entirely from the successful implementation of the government's family-planning pro- gramme. A willingness or otherwise to accept the government's fertility norms was clearly influenced by the socio-economic circumstances of individual families, and levels of fertility continue to reflect levels of development.

Major regional differences exist in the extent and timing of the fertility decline. By 1981 the TFR had declined to well below replacement level in the major municipalities (Shanghai 1.3, Beijing and Tianjin 1.6) and in the more developed provinces (Liaoning and Jilin 1.8) but remained high in many of the less developed provinces of the interior, with rates of 3.8–4.4 in Guizhou, Guangxi, Ningxia, Qinghai, Xinjiang and Yunnan (Table 5.2).

Within provinces, substantial variations exist in levels of fertility between the urban and rural populations. In the 1950s the level of fertility in urban China was about 10 per cent below that of the countryside, and much of the difference could be accounted for by later marriage among the urban population. By the mid-1960s urban fertility had declined to about half that of the rural population, and that ratio has been maintained ever since. In 1980 the urban TFR was down to 1.15 in contrast to the 2.48 of the rural population. Graphs of age-specific fertility rates for urban and rural women in 1970 and 1980 (see Figure 5.4b on p. 108) highlight both the dramatic decline in fertility and the marked rural–urban differential. By 1980 rural fertility had been reduced to below the level achieved by the urban population a decade earlier and such a dramatic and rapid decline of fertility in so large and poor a country as China is quite remakable.

THE 1980s AND THE ONE-CHILD POLICY

Despite the demographic success of the 1970s the leadership decided that an even more stringent population-control policy was required. In 1979 the National People's Congress put forward the policy of 'one couple, one child'. The minister in charge of family planning summarised China's objectives as 'the control of population quantity, the improvement of population quality and the mutual adaption of population and socio-economic development'. In pursuit of these, population policy was incorporated into the country's new 1982 Constitution.

Thus, for the first time since 1949, family planning received the legal backing of the state. Article 25 of the Constitution decrees that 'the state shall carry out family planning in order to bring population growth into line with the plans for economic and social development', and Article 49 stipulates that 'both husband and wife have the duty to carry out family planning'. Various factors combined to persuade the government of the need for a one-child policy, not least the cost of catering for a large and expanding population, the inbuilt momentum for rapid population growth arising from the baby-boom of the 1962–73, and various predictions regarding the optimum population over the next century.

Response to crisis

Disappointment with past economic performance and apprehension about demographic potential precipitated the one-child policy of 1979. The two-child prescription of the 'later, longer, fewer' policy appeared incapable of containing the demographic pressure that would arise as the baby-boom of 1962–73 reached the age of marriage and child-bearing in the 1980s and 1990s. Thus the State Council deemed it 'necessary to launch a crash programme over the coming twenty or thirty years calling on each couple, except those in minority nationality areas, to have a single child. ... Our aim is to strive to limit the population to a maximum of 1,200 million by the end of the century.' If the two-child policy was allowed to continue, the projected population increase, estimated at 15–20 million a year in the late-1980s, seemed likely to jeopardise the economic goals of the country.

If, as the government believed, the problems of feeding, clothing, housing, educating and employing the population are all to be traced to a common problem – too many people – then the solution appeared obvious: produce fewer people. In pursuit of that belief came the one-child policy aimed at reducing not only the growth rate but eventually the size of the population. In the late-1970s a series of population projections were

Table 5.5 Population projections 1980–2080 (in millions)

	Average number of children born to each woman				
	3.0	2.5	2.0	1.5	1.0
1980	986	983	978	978	978
1990	1,180	1,136	1,088	1,048	1,021
2000	1,420	1,323	1,222	1,130	1,050
2010	1,636	1,472	1,310	1,167	1,044
2020	1,915	1,641	1,388	1,175	1,003
2030	2,233	1,828	1,469	1,177	951
2040	2,562	1,995	1,518	1,150	878
2050	2,949	2,164	1,542	1,087	771
2060	3,338	2,300	1,517	972	613
2070	3,794	2,448	1,488	853	457
2080	4,308	2,642	1,483	781	370

Source: Song, J.

produced based on various assumptions regarding the average number of children born to each married woman. The outcome of these projections is summarised in Table 5.5.

If the total fertility rate (TFR) remained at 3.0 (the level it reached in 1976–7) then by the year 2000 the population would be 1,420 million, and over the century 1980–2080 it would grow to 4,308 million (roughly equal to the total world population in 1980). In marked contrast, if from 1980 the TFR was kept at 1.5 the population would reach a peak of 1,180 million in 2027 and thereafter decline to 781 million in 2080. An immediate and continued enforcement of a one-child policy would have given a peak population of 1,053 million in 2004 and a total population of only 370 million in 2080.

Optimum population

After examining the growth potential arising from different levels of fertility, the Chinese then endeavoured to define an optimum population for the next century. Based on projected trends of economic development, food requirements, environmental equilibrium and the like, the consensus view (arising from a series of conferences in 1979–80) was that optimum carrying capacity in the next century was a population of around 700 million. Having identified this target population and the time-span appropriate for its achievement (within a century), the required population policies were devised and are currently being implemented.

The objective in 1980 was to reduce dramatically the birth rate and achieve a fertility rate of one child per newly married couple by 1985. Such a rate was to be maintained until the end of the century. From 2000 to 2020 the fertility rate would be gradually increased to 2.16 children per married couple and that rate (the replacement level of fertility) would be maintained thereafter. Such programmed fertility was intended to stabilise China's population at about 700 million by the year 2070.

Sanctions and rewards

To persuade the present generation to conform to the one-child policy the country adopted a system of economic rewards and penalties. The details of the incentive scheme varied from province to province, but the basic offer to parents in 1979 was a 5–10 per cent salary bonus for limiting their families to one child and a 10 per cent salary deduction for those who produced more than two children.

The economic sanctions were to operate for the first fourteen to sixteen years of the child's life. In addition to their salary bonus, parents who signed a one-child pledge gained preferential treatment in terms of access to such scarce commodities as food, housing and health care; while the single child received priority in access to education and employment. Conversely, parents who produced three or more children were to be penalised, not only financially but also in terms of their more limited access to scarce resources and services. An indication of the financial penalty incurred by those who ignored the new regulations can be gauged from the case of a Beijing couple who gave birth to their fourth child in 1980. Over the period 1980–93 the couple are required to pay costs in excess of 3,000 yuan, equivalent to the husband's total earnings for more than four years.

In 1980 the State Council moved away from the 1979 format of a one-child recommendation while showing tolerance for the two-child family, towards the more stringent implementation of the one-child norm. The punishments and penalties previously applicable to those who gave birth to a third child would henceforth be levied on those who produced an unplanned second child. The severest of pressure would be exerted to have all second and subsequent pregnancies aborted. The abortion issue generated a good deal of hostile comment in the outside world. It has been suggested that from 1971 to 1984 more than 101 million abortions were carried out and 53.9 million of those occurred in the five years of the one-child policy from 1980 to 1984.

While the regulations of the 1980s advocate one child and categorically prohibit a third child, there is a very restrictive range of conditions under which a second child is to be permitted. Generally speaking, a second child may be allowed in cases of physical disability, in poor and remote regions, for ethnic minorities, to preserve the family line and, in some situations, when the only child is a girl (see Case Study 5.2 for examples of the regulations from Shanxi province).

Measuring the success rate of the current policy is somewhat difficult, for at the time of the census and fertility survey in 1982 the one-child programme had been running for only three years, giving a very limited

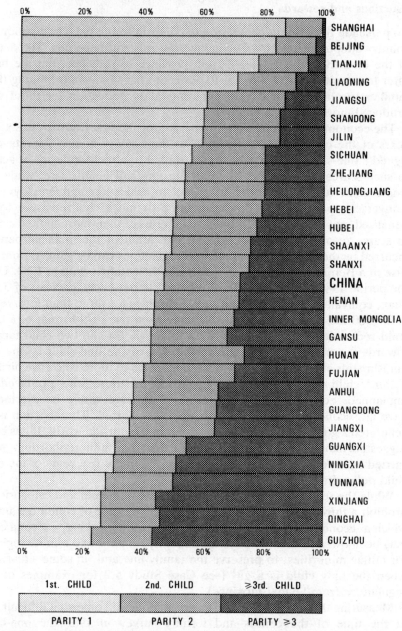

Figure 5.7 Percentage parity distribution of births in 1981 by province

data base from which to judge it. In 1981 first-born children accounted for 47 per cent of all births, the second-born made up 25 per cent and third and higher-order births accounted for 28 per cent. Thus in the early 1980s more than half the births were in excess of the one-child policy.

However, significant progress had been made: between 1970 and 1981 third births declined from 62 per cent to 28 per cent, while first-order births increased from 21 per cent to 47 per cent. By the end of 1984, 28 million married women of reproductive age had been issued with an 'only-child parents' certificate'. In other words, they had one child and had pledged to have no more. Data on the provincial variation in parity order (the proportion of births which are first, second or third and more children) is plotted in Figure 5.7. In 1981 the percentage of first births ranged from 87 per cent in Shanghai to only 24 per cent in Guizhou. In the less-developed interior provinces of Xinjiang, Qinghai, Ningxia, Yunnan and Guizhou around half the births were third or later and almost a quarter were fifth or higher-order births.

NEW SOLUTIONS, NEW PROBLEMS

Solutions to a problem often generate new problems, and there is no doubt that a number have arisen or will arise from the one-child policy. A major long-term problem concerns the ageing of the population and the associated problem of old-age security. Currently, less than 5 per cent of China's population are over 65 years but the pursuit of a one-child policy, even if only partially successful, will totally transform the age structure and in so doing will produce a declining workforce and eventually a very aged population.

Dependents with few supporters

The baby-boom of 1962–73 will eventually produce a retirement bulge in the 2030s. By then, over-65s could constitute more than 25 per cent of the population and thus, within a lifetime, the number of retired will have increased from one in twenty to one in four. Such a high level of old-age dependency is unprecedented even in today's developed countries where the over-65s generally constitute 10–15 per cent of a country's population. Furthermore, if China persists with its present retirement policy, 55 for women, 60 for men, the proportion of retired people among the population of the 2030s and thereafter will be very large indeed.

In China, as in other developing countries, one of the motivations for high fertility is that children are seen as an essential investment for old age, someone to look after you when you can no longer look after yourself. However, in a one-child society there will be too many dependent old and too few working young to fulfil these social obligations. Parents whose only child is a daughter see themselves as particularly at risk, for once married the daughter moves away and joins her husband's family.

Some form of institutional security for old age is needed to replace the traditional practice of 'raising children as a protection against old age'. Yet a state pension is available to only 10–15 per cent of the workforce, being restricted in the main to employees in government jobs (e.g. teachers) and workers in state enterprises. In rural areas less than 1 per cent of the retired personnel receive a pension from the collective welfare fund.

Although there is talk of establishing a rural pension scheme little positive action has been undertaken and the arrival of the agricultural responsibility system has undermined the existing schemes for collective welfare. Funding social security for the elderly will be a colossal under-taking. Already in the developed world the public funding of pension schemes is causing financial difficulties. Yet one of the consequences of the one-child policy is that, in future, each worker will need to support almost twice as many pensioners as is currently the case in the developed world.

Paying for success

Enforcement of the single-child policy through a system of economic incentives and disincentives depends on the ability to finance the scheme. Responsibility for this currently falls on the employers of the husband and wife. Given that agricultural and industrial enterprises command vastly differing levels of welfare funding there is considerable variation in the rewards offered to single-child families. A 1980 survey showed that some of the more profitable enterprises were paying bonuses of 100 yuan and more, others offered woollen blankets, brocade sheets, radios and the like, while in the poorer establishments bonuses were as low as 10–20 yuan or a couple of towels, a thermos flask, a few toys, a bowl, or nothing at all. The equitable implementation of the policy would appear to require a nationally funded system of one-child benefits, but as yet no such scheme is on offer. The uncertainty over funding, especially if very large numbers of parents apply for a one-child certificate, may prove to be the weakest aspect of the incentive programme.

Female infanticide

The clash between the traditional preference for sons and the current one-child policy has unwittingly precipitated a resurgence of female infanticide and the maltreatment of women who give birth to daughters. Such acts of despair were extensively documented in the Chinese press in the early 1980s. In Huaiyuan county in Anhui province, for example, there was a ratio of 133 males to 100 females among the 24,255 births recorded in 1980–81. In the eastern district of Hefei city it was reported that over fifty baby girls were drowned within two months of the implementation of the one-child policy. Attempts are being made to counteract the traditional preference for sons; posters normally show the only child as a girl and several provinces offer larger economic incentives when the only child is a girl. Anhui, for example, offers 5 yuan per month for a son but 6 yuan for a daughter.

Single-child syndrome

Anxieties have been expressed that the single child might acquire undesirable social characteristics, becoming selfish, unco-operative, and maladjusted. There has been much talk about the 'spoilt brat syndrome', and the Chinese have coined the pejorative expression 'little emperor' to descibe the over-indulged single child. The few studies so far undertaken suggest that China's single children have scored higher in verbal skills and intellectual development but less well in measures of personality traits and social behaviour. Substantial research will need to be undertaken to understand the impact of the one-child family both on the child and on society.

CONCLUSION

The pace of demographic modernisation has been most impressive. China currently enjoys advanced levels of demographic development (low death-rate, low birth-rate, long life-expectancy, etc.) at a very early stage of economic development. In terms of a demographic quality-of-life index which combines infant mortality, life-expectancy and literacy, and ranges from zero to 100, China scores 77 in contrast to the 43 of India, the 94 of the USA and a world average of 65. Whether the comparative framework is global or Asian (Table 5.6) it is evident that China needs to be compared with the more developed rather than the less developed countries. Level of fertility and life-expectancy are completely out of character with its low level of economic development. Life-expectancy is fifteen to twenty years greater than would be predicted from the country's level of per capita income. Put another way, to achieve China's current level of life-expectancy one would anticipate a per capita income almost ten times greater than that which currently prevails.

Much of the decline in fertility has been orchestrated by a government with a firm belief in its ability to introduce and maintain the momentum for social change. The one-child policy is certainly one of the most ambitious exercises in social engineering which the world has so far seen. Doubts and reservations can be expressed on many aspects of it. The central argument is that the country's large and rapidly growing population has had a very detrimental impact on the country's social and economic development. But

Table 5.6 Indicators of demographic and economic development in selected Asian countries

	China	Bangladesh	Hong Kong	India	Indonesia	Pakistan	Philippines	S. Korea
1. Population	1,008	93	5	717	153	87	51	39
2. Per capita GNP	310	140	5,340	260	580	380	820	1,910
3. % agric. labour	69	74	3	71	58	57	46	34
4. % contraception	69	19	72	28	53	–	36	54
5. Birth rate	19	47	18	34	34	42	31	23
6. TFR	2.3	6.3	2.1	4.8	4.3	5.8	4.2	2.7
7. Death rate	7	17	5	13	13	15	7	6
8. Infant mortality	67?	133	10	94	102	121	51	32
9. Life-expectancy	67	48	75	55	53	50	64	67
10. Food intake	2,526	1,952	2,920	1,906	2,342	2,313	2,318	2,931

Source: World Bank, *World Development Report, 1984.*
Notes: Data is for 1982 unless otherwise stated.
1. Total population in millions.
2. Per capita Gross National Product, in US dollars.
3. Percentage of labour force engaged in agriculture, 1980.
4. Percentage of married women of child-bearing age using modern contraceptives.
5. Crude birth rate per 1,000 population.
6. Total fertlity rate is roughly equivalent to the number of children born to each woman in the age group 15–49.
7. Crude death rate per 1,000 population.
8. Deaths of infants under 1 year of age per 1,000 live births.
9. Life-expectancy at birth in years.
10. Kcals per person per day.

is this really the case? In China, as elsewhere in the world, the impact of population growth on economic performance remains poorly understood. No clear-cut conclusions can be drawn about the adverse impact of rapid population growth on social and economic development. One must certainly caution against the unqualified belief that a reduction in the rate of population growth offers a panacea for China's future development.

Past economic problems have arisen from a wide variety of causes, not least major fluctuations in economic policy, problems of management and investment priorities and, at times, a lack of motivation among the workforce. The evidence since 1978 suggests that it is the pursuit of different economic, not demographic, policies which has rapidly improved the prosperity of the people. Furthermore, is it reasonable to expect a government which has shown repeated and substantive changes of policy to maintain a long-term commitment to its population policy in the future?

It should also be remembered that birth control is not just a technical and demographic issue. It raises a wide range of moral and ethical questions relating to both the policy itself and also the means by which it has been implemented. The government nevertheless persuaded itself of both the need and the right to intervene in the matter of procreation and did so with dramatic success. Fertility was halved in the 1970s. At the beginning of that decade couples generally married early, did not practise contraception and achieved high levels of fertility. By the end of the decade the combination of late marriage, the adoption of modern forms of contraception by almost 70 per cent of couples and the considerable use of induced abortion meant that fertility was reduced to almost replacement level. The total fertility rate had declined from 6.448 in 1968 to 2.238 in 1980.

Looking to the future it may well be that the new economic reforms will, by increasing the pace of social and economic development, generate a desire for fewer children and in doing so will narrow the gap between the goals of policy-makers and the wishes of people. Finally, a new phase in the population drama may be just beginning. A recent projection suggests that by the end of the century China's population will be split 50 : 50 between the rural and urban areas, with about 600 million in each sector. Such a projection implies that over the period 1984–2000 the rural population will decline by about 100 million while the urban population will increase by 270 million, an increase of over 80 per cent in the level of urbanisation. The problems posed by such a dramatic residential transformation should continue to tax the ingenuity and resourcefulness of the Chinese government well into the next century (see Chapter 8).

REFERENCES

(1969) *Selected Works of Mao Tse-tung*, vol. 4; Beijing.

FURTHER READING

Aird, J. S. (1981) 'Fertility decline in China' in N. Eberstadt (ed.) *Fertility Decline in the Less Developed Countries*, New York.

Banister, J. (1987) *China's Changing Population*, California: Stanford University Press.

China's Financial & Economic Publishing House (1988) *New China's Population*, London and New York: Macmillan Publishing Co.

Coale, A.J. (1984) *Rapid Population Change in China, 1952-82*, Washington: National Academy of Sciences.

Croll, E. *et al.* (1985) *China's One-Child Family Policy*, London: Macmillan.

Kane, P. (1988) *Famine in China 1959-61: Demographic and Social Implications*, London: Macmillan.

Kane, P. (1987) *China's Second Billion*, Harmondsworth: Penguin.

Li, Chengrui (1987) *A Census of One Billion People*, Boulder: Westview Press.

Population Census Office of the State Council (1988) *The Population Atlas of China*, Oxford: Oxford University Press.

World Bank (1983) *China: Socialist Economic Development*, vol. 3: *Population, Health, Nutrition and Education*, Washington: World Bank.

UPDATE

As well as the sources suggested for Chapter 1, the *Journal of the International Planned Parenthood Federation* (IPPF) carries information on China from time to time. The February 1989 issue was a special one on China and is available separately from the organisation's office at Regent's College, Inner Circle, Regent's Park, London NW1 4NS.

Case Study 5.1

MIGRATION

John Jowett

International migration

Since 1949, levels of international migration to and from China have been very low. There has been a continuous flow of migrants, both legally and illegally, from the mainland into Hong Kong, with an estimated half a million crossing the border in the peak period 1977–82. Smaller emigration streams include the thousands of Tibetan refugees who left for India after the unsuccessful uprising in Tibet in 1959–60 and the 50,000 or so Khazaks who crossed the border from Xinjiang to the USSR in 1962 (see Chapter eleven). The two important phases of immigration involve ethnic Chinese from Indonesia and Vietnam. In 1960, following anti-Chinese riots and massacres in Indonesia, some 90,000 Chinese refugees went to the mainland and a further 5,000 in 1966. In the late-1970s some 265,000 ethnic Chinese crossed from Vietnam or left by boat and were settled in south China (see Chapter eleven). The majority of these refugees were pushed across the Sino–Vietnamese border in 1978.

Inter-provincial migration

Lack of data has greatly restricted studies on internal migration. Recent estimates suggest that some 30 million people were involved in inter-provincial migration in the period 1949–84. Such a figure underestimates the true extent of migration since it excludes military personnel and temporary migrants. In China, as elsewhere in the world, migrants tend to be age- and sex-specific, with the pioneering phase of migration domi-nanted by young, unmarried men. While most of the migration involving skilled and military personnel is state-controlled, much of the rural–rural migration has been of a spontaneous nature, what the Chinese refer to as self-drifting.

The reasons for migration have been many and varied. The movement of military personnel into sensitive border provinces in support of ter-ritorial disputes has generated migrations into Guangxi and Yunnan (Sino–Vietnamese dispute), Tibet (Sino–Indian), Xinjiang, Gansu, Inner Mongolia and Heilongjiang (Sino–Soviet) and to Fujian in support of the Taiwan claim. Resettlement programmes have attempted to move popula-tion from the densely populated provinces of the east to the sparsely populated provinces in the west and northeast. This redistribution was partly an end in itself, but it was also in support of economic and social development in these underdeveloped regions.

The east-to-west migrations were also designed to alter the ethnic composition of populations in the west. In the case of Xinjiang the proportion of Han Chinese rose from less than 10 per cent in 1949 to 40 per cent in 1982. By placing the incoming Han Chinese in the key administra-tive and political positions, Beijing maintains strong central control in these outlying and strategically important provinces. Political prisoners, intellectual exiles and criminals sent to remote labour camps have also formed part of the inter-provincial migration flows.

While much of the rural–urban migration has taken place within provincial borders, some movements have been between provinces. During the major de-urbanisation movements of 1966–78, when educated urban youths were sent 'up to the mountains and down to the countryside', hundreds of thousands were transferred from eastern cities to the border provinces of Heilongjiang, Inner Mongolia and Yunnan (see Chapter eight).

Over the period 1953–82 inter-provincial migration flows have significantly increased the population in the northern provinces of Heilongjiang and Inner Mongolia, and in Ningxia, Qinghai and Xinjiang in the west. Over a 30-year period the populations in these provinces have grown at an average of 3.0–3.5 per cent per year, much more than the rate of natural increase. The major source area for the migrants has been eastern and central China; in provincial terms, Hebei, Henan, Shandong, Anhui, Jiangsu, Shanghai, Hunan and Sichuan.

Inter-provincial migrants have supported major industrial developments, not least in the oil sector, and have also assisted with major projects for land reclamation and the development of state farms in the sparsely populated semi-arid provinces. Thus, among the major pioneering migration streams into Qinghai the one in 1956 involved 69,700 people from Shandong, Henan and Hebei. A second migration wave in 1958 consisted of 44,300 people from Henan who were located on twenty-nine state farms in six of the provinces' autonomous prefectures. A third phase involved 14,000 young people and demobilised military personnel who were recruited from Shandong to work on the Gurmu state farm. In Qinghai, as in several other interior and border provinces, between a third and a quarter of the population are migrants. The demographic, social and economic development of these regions has been strongly influenced by the migration process.

Rural–urban migration

During the First Five Year Plan (1953–7), when rural–urban migration was not greatly restricted, millions of peasants moved into the towns and cities seeking job opportunities during an intensive phase of economic reconstruction. Between 1949 and 1957 the urban population increased by about 60 per cent (from 49.0 to 82.1 million) while the rural population grew by only 13 per cent. During the Great Leap Forward (GLF) the pace of rural–urban migration quickened still further as China went in search of 'instant', labour-intensive, industrialisation, and the urban population soared to 109.5 million in 1960 (see Table 8.1, adjusted figures).

While in most developing countries rural–urban migration is dominated by a one-way flow from the countryside to the towns and cities, in China this is not the case. With the collapse of the GLF and the severe famine in 1960–1 the rural–urban migration flows of the 1950s were rapidly reversed. The 'surplus' population was moved out of the towns and cities and returned to the countryside. Some measure of the extent of the reverse migration can be gauged from the fact that the urban population declined by 11 million between 1960 and 1964. To limit the size and growth rate of the urban population in the 1960s and 1970s, the government imposed strict controls on rural–urban migration.

It also initiated a policy of sending educated urban youths to the countryside (the so-called 'rustication programme') as an intervention in support of urban–rural migration. Between 1968 and 1980 some 15–20 million urban youths were rusticated. Much of the urban–rural migration involved in the programme occurred within provincial boundaries. For some of the 'sent-down' youths the transfer was a temporary, two- or three-year assignment, but for others it represented a permanent change of residence and lifestyle.

The available data on rural–urban migration relates to the very large cities such as Shanghai, Beijing and Tianjin. In these cases the volume of migration in and out of the cities in any one year tends to be very large relative to the net movement. In the case of Tianjin the period 1950–80 saw 3.14 million people move into the city and 2.79 million move out, resulting in a net in-migration of 350,000. From 1949–83 the migration flows in and out of Shanghai resulted in a net loss of a million people. This, combined with the low level of natural increase over the past twenty-five years, has kept the population of the city proper at or below 6.5 million since 1957. The changing direction of net migration between the cities and the countryside is well illustrated by Tianjin's experience. With a net in-migration of almost half a million in the 1950s, this was cancelled out by a similar reverse flow in the 1960s.

Recent developments

The political and economic reforms of the post-1978 period have had two important influences on migration. First, with the relaxation of controls on the movement of population, there has been a substantial increase in the number of temporary migrants. Surplus rural manpower is now allowed to migrate freely without permanent registration. Skilled workers, in building trades and other occupations, now travel widely in China. For several months, or even a few years, these temporary migrants work outside their home province. Second, the east–west and urban–rural migration flows of the Maoist era have been somewhat reversed. In the years immediately following Mao's death the new leadership initially insisted that the sent-down youths remain in their assigned locations and that new graduates from schools and colleges continued to be rusticated. However in 1978–9, in a surprising show of civil disobedience, newly assigned youths refused to leave the cities and former sent-down youths returned home. At this point the government relented and allowed millions of former exiles (students, cadres, intellectuals) to return home to the more developed regions, in particular to the urban areas of eastern China. The late-1970s and early-1980s saw a dramatic increase in the rate of migration. In Shanghai the net in-migration in 1979 alone was 264,800. However, official government policy is attempting to divert the rising wave of rural–urban migration away from the big cities and towards the development of medium-sized towns.

Case Study 5.2

FAMILY PLANNING REGULATIONS FOR SHANXI PROVINCE

John Jowett

Some indication of the incentives and disincentives used in support of the one-child policy can be gained from the following abstract of the planned-parenthood regulations for Shanxi province. The current format is not the one used in the original published in December 1982.

Except in 'special circumstances' and cases of 'real difficulties', when the birth of a second child is approved, couples are allowed only one child. Under no circumstances is the birth of a third child permitted. 'Special circumstances' are as follows:

1 The first child suffers from a non-hereditary disease or disability and is unable to develop into a normal working adult.
2 One partner of a remarried couple has children from a previous marriage and the other does not.
3 Pregnancy occurs after a couple, who have been childless for many years and have been certified as sterile, adopt a child.
4 Both husband and wife are of minority nationality.
5 The couple are returned Overseas Chinese.

'Real difficulties' are as follows:

1 People living in sparsely populated mountain villages with poor natural resources and inadequate communications.
2 Both husband and wife are only children.
3 Only one of three (or more) brothers is able to reproduce.
4 Only one son has been born into a family for three generations in a row.
5 An only daughter marries uxorically (the husband joins the wife's family).
6 The only son of a martyr.
7 One spouse has a first-degree deformity.

'Special circumstances' are applicable to the whole province. 'Real difficulties' are only applicable to the countryside. An interval of at least four years is required between the birth of the first and second child.

Late marriage and delayed child-bearing

Couples who marry late (husband and wife are at least 23 and 25 years old) qualify for an additional fifteen days of marriage leave. Women who qualify for deferred child-bearing are entitled to a minimum of 100 days of paid maternity leave.

Incentives for the one-child family

1 Income supplement: (a) in urban areas (state cadres, staff, workers and urban residents) a monthly health subsidy of 5 yuan until the child is 14 years old; (b) in rural areas couples shall be given 5 work points per month or the cash equivalent until the child is 14 years old. Lowered production quotas may also be adopted.
2 Educational benefit: priority in admission to nurseries and kindergarten, and where conditions permit admission entirely or partially free of charge; priority in access to higher education.
3 Medical benefit: priority in registration, examination and hospitalisation for sick children.
4 Housing benefit: priority in housing assignment.
5 Food benefit: in rural areas the only child receives an adult grain ration.
6 Old-age security: in rural areas the parents of an only child who is not around to look after them in their old age are guaranteed a standard of living no lower than the average in the local community.
7 Rewards will be paid to individuals and units which make outstanding contributions to birth-control work.

Disincentives to exceeding one child

1 Return of benefits: couples who have an unplanned second child after being rewarded for only having one must return, within a limited period, all benefits so far received.
2 Income deductions: (i) during pregnancy: women whose pregnancies are not covered by the plan should be persuaded to undertake timely *remedial measures* (a euphemism for abortion). Those who ignore this advice are subject to the following penalties:
   (a) in urban areas (cadres, staff and workers) couples will suffer a 20 per cent deduction from their combined annual basic pay for a second pregnancy and 30 per cent for a third;
   (b) in rural areas (rural commune members) 20 per cent will be deducted from the couples' annual income for a second pregnancy and 30 per cent for a third, or other corresponding economic sanctions shall be taken;
If *remedial measures* are taken all the deducted monies will be refunded.
3 Income deductions: (ii) after the birth:
   (a) in urban areas, for the unplanned birth of a second child, both husband and wife will have 15 per cent of their monthly pay deducted for seven years and both will be deprived of one future pay

131

increase. In the case of a third child 10 per cent of the monthly pay of both husband and wife will be deducted for a period of fourteen years, the couple will be demoted by one grade on the pay scale and they will be deprived of one future pay increase;

(b) in rural areas, for an unplanned second birth, both husband and wife will have 10 per cent of their total annual work points deducted for a period of seven years. Alternatively they will be allocated less responsible plots or assigned higher output quotas; one of their private plots will be called back or the proportion of their produce to be retained by the collective will be increased. In addition, 10 per cent of their total annual income will be deducted for two years, or other corresponding economic sanctions will be taken. In the case of a third birth the wording is exactly the same, with the penalties increased to 10 per cent of their total workpoints for fourteen years and on the alternative scale 20 per cent of their total annual income for three years;

4 Medical costs: the couple will be deprived of all medical, material and other benefits granted to a planned birth; maternity leave will be granted but without pay.

5 Housing costs: in urban areas no additional housing space will be granted and in rural areas no additional land for housing construction will be made available and no additional private plot will be granted.

6 Hardship costs: no subsidies will be paid for hardship arising from having extra children.

7 Employment costs: couples who give birth to unplanned children shall not be recruited for five years. If they are contract workers, workers employed according to oral agreements, temporary workers or recruited workers from rural areas, they will be dismissed.

8 Those who stubbornly refuse to give birth according to plan and whose offences are serious shall be given severe disciplinary punishment in addition to the economic sanctions.

9 Units and departments which fail to fulfil their planned parenthood tasks, do not implement planned parenthood policies and regulations or do not register the birth of all children will be subject to disciplinary punishment or economic sanctions.

Chapter six

Rural China: old problems and new solutions

Frank Leeming and Simon Powell

China is a rural country, and the achievements and problems of the rural community lie at the heart of any study of its geography and society. The central achievement of rural China is well known but not always recognised in proper perspective – it is the maintenance of about one-fifth of the world's population in food and clothing with only 7 per cent of the world's arable land. This achievement is made possible by intensive and frugal management of a body of arable land of which about half is of very good quality, and of which the greater part has been improved by human action.

This improvement and intensification of use did not begin with Liberation in 1949. These processes have a long history bound up with the historic growth of the vast Chinese community. As early as 1850, the population of the then Chinese Empire was around 400 million. Population has risen to more than double this figure since Liberation, with the community increasingly dependent on the arable land, since imports of food and cotton remain small. Meanwhile, social organisation has run through a remarkable course of change, involving an advanced socialist system in the phase 1958–78, and the purposeful demolition of many features of that system since 1978.

RURAL RESOURCES AND PATTERNS OF LAND USE

In terms of rural resources China falls into two main regions of roughly equal size: the seaboard, with its adjacent areas, extending far inland, and the central Asian mountain ranges, plateaux and basins. These are separated by the heavy line in the land resources map (Figure 6.1). The eastern region is essentially that of 'China proper' (see Figure 3.2 on p. 64). To a substantial degree, the 'rock and desert' classification in the map key indicates areas lying within this central Asia region. Much of the 'prairie and grassland' land category also lies within this region, which, as various studies in this volume also show, is a world which displays marked differentiation from Inner China in both traditional and modern times.

Land resources in the eastern half of China differ greatly in character

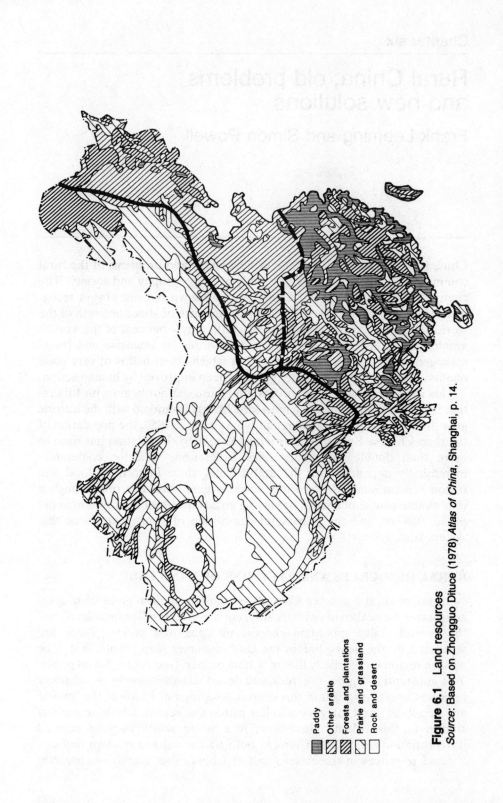

Figure 6.1 Land resources
Source: Based on Zhongguo Dituce (1978) *Atlas of China*, Shanghai, p. 14.

Paddy

Other arable

Forests and plantations

Prairie and grassland

Rock and desert

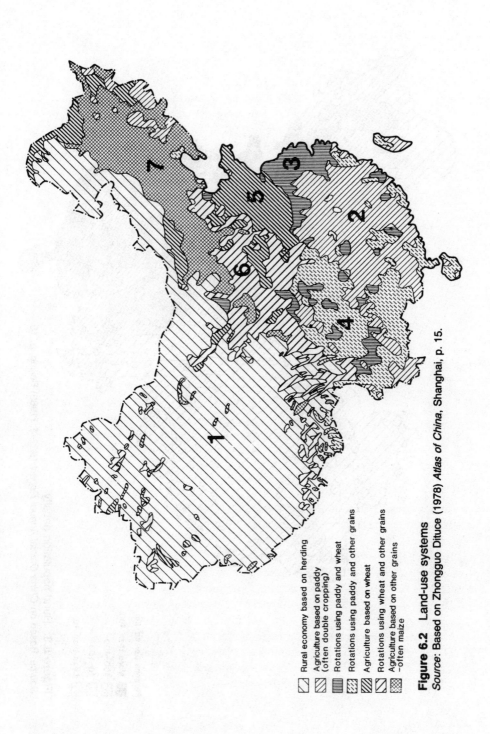

☐ Rural economy based on herding

☒ Agriculture based on paddy
   (often double cropping)

☰ Rotations using paddy and wheat

☒ Rotations using paddy and other grains

☒ Agriculture based on wheat

☒ Rotations using wheat and other grains

☒ Agriculture based on other grains
   – often maize

Figure 6.2 Land-use systems
Source: Based on Zhongguo Dituce (1978) *Atlas of China*, Shanghai, p. 15.

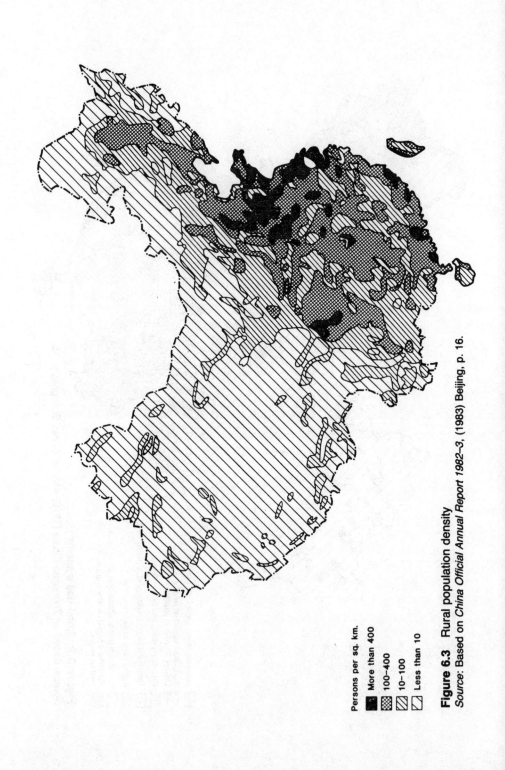

Persons per sq. km.

■ More than 400

▨ 100–400

▧ 10–100

◇ Less than 10

Figure 6.3 Rural population density
Source: Based on China Official Annual Report 1982–3, (1983) Beijing, p. 16.

from region to region – whether mountain or plain, and whether north or south. Differentiation between north and south is fundamental. Southern China is generally well-supplied with rain, and experiences a mild and relatively short winter. Agricultural resource in northern China suffers from shortage and excessive seasonality of rainfall (less so in the north-east), and from harsh and relatively prolonged winters. The boundary zone between south and north in these respects (roughly the line of the Qinling scarp and Huai River, marked by a broken heavy line in Figure 6.1) represents a critical part of a transition zone which extends through most of central China.

It is critical not only because of climatic differences, but also because of differences in soils. To the north of the Qinling–Huai line, soils are loose and calcareous and, although fertile, generally dry; to the south, soils are generally acid, and in the lowlands, mixed and sticky. Soils in the south are adapted to the creation of paddy fields; those in the north are not. It is this, rather than lack of warmth in Summer, which limits the broad distribution of rice farming to the south, as shown in the land-use systems map (Figure 6.2).

Farmland in the northern half of eastern China is usually occupied by food crops other than paddy: wheat, maize, millet, barley, potatoes. The detail in the land resources map shows this distinction, together with the distribution of forests, mainly in the mountains. Non-paddy farmland in southern China also lies mainly in mountain areas. It will be observed that the land-use pattern is much more complex in the topographically varied environments of southern China than in the north, where topography is dominated by level plains or elevated plateaux (see Chapter four).

Regions in rural production

The land-use systems map (Figure 6.2) advances this analysis a stage further. It represents differences among production systems based on a particular food crop. Food crops are the heart of the matter in Chinese agriculture; relatively little effort is devoted to animal feedstuffs, grazing or industrial crops by contrast with Europe or North America. The overwhelming majority of farmland is used to produce food, and in almost all areas the overwhelming majority of food output is used to provide subsistence for local people. Rural population distribution and density (Figure 6.3) very closely match the agricultural potential and productivity of the various regions distinguished here. The regions are numbered, and will be discussed in order. Tables 6.1 and 6.2 use the same numbers and represent analysis of the same regions.

1 Central Asian region

Figure 6.2 suggests at once the separation of the central Asian half of China from the eastern half. The vast central Asian region is not primarily

Table 6.1 Rural land-use regions of China

	1 Central Asia – Xinjiang	2 South and southeast	3 Lower Yangzi	4 South west	5 North China Plain	6 Loess plateau	7 Northeast and north
Typical precipitation (millimetres)	100	1,500	1,250	1,250	600	400	600
Seasonality of precipitation (months with more than 10% of total)	June, July, Aug., Sept.	April, May, June, July, Aug., Sept.	May, June, July, Aug.	June, July, Aug., Sept.	June, July, Aug.	June, July, Aug., Sept.	June, July, Aug., Sept.
Typical temp. range (°C, monthly averages):							
January	−12	10	3	8	−3	−5	−15
July	25	28	28	19	26	24	23
Frost-free days	200	300	240	250	200	180	150
Typical farm system – grain crops	1 crop per year: maize, wheat, rice, barley.	2 crops per year: rice, maize, Winter wheat.	2 crops per year: rice, Winter wheat.	2 crops per year: or 3 crops in 2 years: rice, maize, Winter wheat.	3 crops in 2 years Winter wheat, maize, millet	1 crop per year: Spring wheat, millet, potatoes.	1 crop per year: Spring wheat, maize.
Typical non-grain crops	Tobacco, cotton.	Tea, mulberry, oranges, forests.	Cotton, rape-seed	Tobacco, tea, forests.	Cotton, tobacco, fruit.	Cotton, rape-seed	Soya beans fruit, sugarbeet forests.
Proportion of arable area to total area (%)	1	15	35	10	40	8	20
Proportion of China's total arable area	9	25	10	5	23	12	16

Note: These details are intended to illustrate the characteristics of China's land-use regions. Great differences exist among regions according to precipitation, temperatures, frost-free days, and so forth. Length of the Summer relates closely to the possibility of double-cropping. Seasonality of precipitation relates closely to environmental stability of land under agriculture. Low proportions of land area under agriculture depend partly upon low precipitation (in central Asia); partly upon mountain conditions (in south and southeast; also southwest).

arable, though increasing proportions of the region's food supplies now come from the limited, but often highly productive, oases and valley bottoms where agriculture is practised. Population throughout this region is sparse (Figure 6.3). Summers are usually hot, except in Tibet, though

Table 6.2 Provincial figures illustrating development in the seven agricultural regions (figures for 1984 except where indicated)

	China (totals/averages)	Xinjiang (northwest)	Guangdong (South and southeast)	Zhejiang (lower Yangzi) (southeast)	Yunnan (southwest)	Henan (North China Plain)	Gansu (loess plateau)	Jilin (northeast) and north
Population (millions)	1,035	13	62	40	34	76	20	23
Rural workforce (millions) (1982)	339	3	21	16	13	27	6	4
Sown area (million hectares) (1982)	145	3	6	5	4	11	3	4
Rural workforce per hectare of sown area	2.3	1.0	3.5	5.3	3.3	2.5	2.0	1.0
Grain yields (tonnes per hectare of sown area)	3.6	2.5	4.1	5.2	2.9	3.2	1.9	4.7
Rural electricity use kWh per hectare of sown area	320	200	417	640	152	236	333	300
Chemical fertiliser (kg per hectare of sown area)	121	67	166	153	97	123	59	123
Percentages of rural outputs from local industry (by value)	15	4	8	28	7	13	5	9
Gross values of rural outputs per rural worker (RMB)	1,107	1,733	1,019	1,281	641	893	717	2,450
Gross values of rural outputs per hectare of sown areas (RMB)	1,511	1,733	3,567	4,100	2,084	2,191	1,433	2,450

Source: Tables in *Statistical Yearbook of China, 1985* (and *1983*)

Note: In this table, individual provinces have been selected to represent the seven agricultural regions. The figures given in the columns are those for the provinces named, together with those for the whole of China in the first column. Some of the categories for which figures are given may be taken as indices of advancement (e.g. rural electricity use); some as indices of prosperity (e.g. gross values); and some as indices of crowding and intensification (e.g. rural workforce).

winters are very cold. Grains grown are wheat, barley and maize, but may also include rice. Traditional livelihood systems in most of the central Asia region were those based on animal herding (pastoralism), often among peoples of nomadic habit. Nomadic peoples are increasingly settled, but herding remains important.

2 The south and southeast

The region which contrasts most strikingly with the central Asia region is the south and southeast, where paddy is dominant. Favoured areas here, such as the plains of Guangdong on the south coast, grow two crops of rice in the year, the first sown in seedbeds in March, planted out in May and harvested in July, the second sown, also in seedbeds, in May, planted out in August, and harvested in November. Rural populations in this kind of area reach very high densities indeed, locally of the order of 2,000 per sq. km of arable, or even higher (Figure 6.3). Full double-cropping is a system which requires an abundant labour supply at busy times of the year, as well as one which can support a dense population. Over most of the area shown on the map as 'agriculture based on paddy' full double-cropping is not yet routine, but it is usually possible, and is one of the most important means by which outputs can be increased.

3 Lower Yangzi region

Another part of the south which shares the quality of very dense development is the lower Yangzi, where, as the map shows, the characteristic agricultural practice is double-cropping. This system uses a Summer crop of rice (sown in April, planted out in June, harvested in September) against a Winter crop of wheat (sown in October, harvested in May). Here, too, rural population is exceptionally dense – but of course in most parts of rural China populations are now approaching double their densities of forty years ago.

4 Southwestern China

The map also distinguishes a large, discontinuous, and very varied region in southwest China, which uses rotations with paddy and dry grains (usually maize, potatoes or wheat). The climate here is cooler than most of the south, especially in Summer, due to high altitudes. Chinese writers call it 'perpetual Spring'. Here, population density varies greatly from sparse to fairly dense, and double-cropping is not routine. Instead, three crops may be grown in two years. In mountainous areas, paddy fields can often be created in valley bottoms and even on hillside shoulders, but neither topography nor water supply can be adapted to build them everywhere. In some areas, including the southwest, a good deal of farming takes place on hillside fields which cannot be used for paddy.

5 North China Plain

Turning to the northern half of eastern China, there is again a distinction evident in the map, between lowland and upland. The North China Plain's grain output is primarily wheat. Full double-cropping is difficult in the plain, partly because of the shorter Summer, partly because wheat is usually a Winter crop (sown in October and harvested in May), but the most favoured Summer crop, which is cotton, requires a long Summer and cannot be united with a Winter crop. Summer crops of maize or millet are, however, also possible. The net result is that in the North China Plain three crops may be taken in two years, combined with one short Winter fallow or a green fertiliser crop. The North China Plain's dense rural population occupies habitats less favoured then those of the south. Rural labour is often in surplus in the plain, sometimes to the extent of 30 per cent or even more.

6 The loess plateau

To the west lie the open slopes and deep gulleys of the loess plateau, together with its mountain peripheries. The loess plateau is a weak environment and suffers from drought conjoined with serious erosion which results from the badly needed but sometimes torrential rainfall (see Chapter four). Here, wheat is allied with millet and maize as staple crops. One crop per year is the rule. Populations are not dense except in the rich valleys of the southeast of the region, but in some areas subsistence is barely adequate.

7 Northeast and north China

These are two different environments, with serious desiccation (and in places desertification) in the Mongolian plateaux lying to the north of the loess areas; and in the northeast a relatively favoured environment with adequate water, its main weakness the intense cold of winter.

In broad terms, these two environments share one farm system. The staples of subsistence are dry grains: wheat and millet in the Mongolian plateaux; maize and wheat in the northeast, together with potatoes. All these are Summer crops. The northeast is a prosperous region whose settlement by Chinese was delayed until the present century; its agriculture is more large-scale and more mechanised than that of most other parts of the country.

FARM SYSTEMS AND INTENSIFICATION OF RESOURCE USE

Apart from food-grains, Chinese farmers also grow a wide range of other crops, both for sale and to add variety to diet. In most cases, commercial crops are rotated with grain crops, and provide useful flexibility in local rotations as well as welcome incomes to local communities. Peanuts,

141

rapeseed and sesame seed produce edible oils. Soya beans, used for making bean-curd, a very rich source of vegetable protein, are also useful both in diets and for sale. Important ancillary produce comes from pigs and poultry, which also contribute fertiliser. Fishponds are important, especially in the south.

Additionally there are important and valuable plantation crops, particularly tea and (for feeding silk worms) mulberry. Fruit orchards (apple and pear in the north, orange and litchi in the south), and (in tropical areas) rubber plantations, represent more modern developments. Vegetables and melons are grown everywhere, together with flavourings such as ginger and garlic. All these products represent a very complex range of opportunities for local peasant families, which is traditional and which is now being exploited with much greater energy than ever before. The complete range of possibilities for local farming systems, taking grain and ancillary outputs all into account, is virtually limitless; and as the systems develop in the present phase they will certainly become increasingly complex.

The systems of production now under discussion have very mixed historic origins. Part of their development has been conditioned by the dramatic growth of population since about AD 1400, from an estimated 100 million to the present figure approaching 1,100 million. Throughout this immense growth, the great bulk of supplies of grain and ancilliary foods and clothing fibres (especially cotton) have continued to be drawn from the Chinese countryside. In part, this immense increase has been supported by extensions of the arable area, particularly in the northeast in the present century. But intensification of land use has also played an essential part – double-cropping, increased inputs of labour and natural fertiliser, improved irrigation, and the use of the new crops which came from the Americas after AD 1500 – maize, potatoes, sweet potatoes and peanuts.

The spectrum of modernisation since 1949

The historic intensification which accompanied the population increase of the last five centuries is one context of technical advance promoted by the Communist Party since 1949. Another is the wide range of rural improvements – electricity, diesel pumps, motor transport, farm machinery, chemical fertilisers and pesticides – which has opened up worldwide during the twentieth century. Chinese farming in 1949 was competent, especially in the south, but old-fashioned, dependent upon very heavy labour inputs, and tied to a stagnant social system which took little interest in technical innovation, especially in the north.

In some respects, traditional intensification before 1949 used production systems which were already elaborate, and by traditional standards efficient, particularly in irrigation, the use of natural fertilisers (including pond mud and composted human and animal excreta and vegetation) and the development of transportation by water – all characteristic features of the south. But in relation to other possible improvements, such as chemical fertilisers, pesticides, motor transport, electricity and farm machinery, rural China had virtually nothing to show.

Since 1949 agriculture has received a consistently low proportion of all investment in China, partly because of the state's marked preference for official investment in industry, partly because the poverty of local communities and the instability of ownership and managment due to political changes inhibited local investment. Under the communes, local communities at village level were responsible for their own financial affairs and could not easily see their way to adventurous investment policies.

However, through general media pressure and specific instructions from senior units, many village communities did undertake extensive works in land reconstruction, such as drainage and irrigation works. Others, particularly in the south where suitable mountain sites are common and water abundant, at least in Summer, built local hydro-electric stations (see Case Study 7.2 on p. 200). Experiments were made with local schemes for methane gas supplies (for cooking and light) from bio-gas cesspits (see Chapter seven). Particularly during the Cultural Revolution, the state established many chemical-fertiliser factories at county level, and chemical fertilisers became cheaper and much easier to obtain. Pesticides began to be available in quantity, though this led to pollution problems in some places.

Experience with farm machinery was more complex. Until the 1970s almost all the farm machinery available comprised big tractors not suited to conditions in most parts of China except the northeast; 'walking tractors' (similar to rotovators) suitable for small fields were practically non-existent. Many units which had machines found them very expensive to use and maintain. Morever, where populations are very dense, the labour-saving advantages of farm machines are less evident.

The Maoist official system proclaimed the necessity of all kinds of modern technical innovation but gave little financial support to communities hoping to adopt them, and insisted upon their use mainly in the production of grain, rather than more profitable outputs such as sugar or fruit. Since 1978, with the liberalisation of the production systems under the present regime, modern inputs are increasingly adopted by households as a part of the new rural enterprise systems. This applies particularly to fertilisers and pesticides, but extends also to veterinary care of animals and poultry, and where possible the use of motor transport.

The Maoist landscapes

The course of events in the Communist Party's reconstruction of Chinese institutions after 1949 has been outlined in Chapter one. This reconstruction was both rapid and far-reaching in the countryside. Following upon campaigns of denunciation and expropriation of exploiting landlords, a radical land reform was introduced in 1951 which was reported completed in 1953. Under this reform, ownership of about 40 per cent of the arable land was transferred from landlords to peasant small-holders – in much of China, especially the north, many peasants already owned the land they farmed.

In 1954 and 1955, following campaigns for voluntary association of peasants in mutual-aid teams, a further campaign for the widespread creation of co-operatives nationwide was introduced. In these, peasants pooled their land-holdings which were then managed jointly by a committee, and they also rented their animals and major tools to the co-operative. Then, unexpectedly, in 1956, with Mao Zedong's direct support, the co-operative movement abruptly changed direction and became a movement for universal collectivisation. Under this system the ownership, as well as the management, of arable land and major assets was assumed by the collective and their uses allocated by the managing committee.

In 1958 the collectives were swept into the much broader and more radical movement for the rural People's Communes; all the social ownership forms of the collectives were retained, together with moves towards socialisation of family life, such as the institution of public eating-halls. In addition, the communes were made units of local government, and hence became directly responsible to the state. Some localities also took part in the enthusiasm, in 1958 and the years following, for rural industrialisation. Bad weather followed all this enthusiasm and disruption, and food output declined drastically. In the ensuing famine of 1959–61 death rates rose dramatically in many provinces (see Chapter four). In the early 1960s the breathless race of socialisation slowed down to a crawl, and rural life returned to some degree to its old customs, though the 'big and public' forms (large in scale and social in character) of the communes remained standard.

What are called here the 'Maoist landscapes' took their definitive form in the next phase – in the Cultural Revolution which swept the country in 1966 and whose effects on the countryside, following upon most of the earlier parallel movements which have been mentioned, remained paramount until after 1978. The Cultural Revolution was Mao Zedong's final bid for the creation of a socialist society which should experience its socialism at every level.

'Big and public' institutions were one fundamental component. These were the People's Communes, each managing on average some 3,000 households in about twenty subordinate 'production brigades', each in turn managing about ten 'work teams' (often based in natural villages). The communes and brigades exercised control and management of all rural resources such as land and labour. In these tasks the communes were themselves subordinate to the instructions of the counties and to central government policies. Operating through the annual plans for production in the villages, these controlled in detail the allocation of resources and at the same time the activities of the people.

'Take grain as the key link' was another fundamental. This principle promoted the growing of food-grains on the arable to the virtual exclusion of other crops, and was enforced through the planning system. A third fundamental, never enunciated as policy but abundantly evident from the behaviour of officials in the administration, was the limitation, and even suppression, of outputs produced for sale except those intended for sale to the state commercial agencies. This applied to eggs, poultry, pigs, fish, fruit and vegetables, and predictably led to gross shortages of all these items in town and country alike. It also applied to unauthorised production of such items as clothing, furniture and bricks, even by collective production. It was reinforced by the closure, by administrative action, of some tens of thousands of local markets. The rationale of these restrictions was the suppression of 'capitalist' small business and the prevention of its regeneration.

The rural landscapes which emerged from more than a decade's application of these fundamentals through social control at all levels were heavily committed to grain cultivation and lacked diversification. There was little or no internal commerce except that which could be managed through the notoriously indifferent state commercial organisations. In most areas, food grain supplies for subsistence were adequate, but food of other kinds was scarce. Peasant households usually had very small private plots for growing vegetables for themselves, and pigs were often maintained through them, but these were always under political threat.

One longer-term intention of Maoist policies was to reduce the differences among households, villages, localities and regions; but areas powerfully favoured by nature in various ways, such as the best localities in the south and east, remained relatively prosperous. More important, perhaps, weak localities in the mountains and along the Great Wall periphery in the north, facing the Gobi desert, remained desperately poor. Finally, the Maoist landscapes were official landscapes. Little happened in the villages, in terms of resource allocation and production, which was not officially sanctioned; and that little was subject to energetic official scrutiny and interference.

MANAGEMENT SYSTEMS SINCE 1978

In 1978 peasant enthusiasm for agricultural production was low. Maoist policies of egalitarian distribution – 'eating from the same big pot' – had served to depress incentives within the rural economy. Labour rewards were not linked to agricultural output: remuneration was in the form of work-points which were allotted to peasants on the basis of time spent in the fields rather than the actual work done. There was thus little incentive for an individual to be efficient or hard-working.

Responsibility systems, contracts and family farming

To combat this problem, the state introduced in 1978 a variety of production responsibility systems designed to increase peasant incentive and enthusiasm for production while, at the same time, allowing the state to retain its grip on the production of basic agricultural produce. By the end of 1983 it was clear that the household responsibility system, whereby responsibility for agricultural production was allotted to an individual household, had come to dominate the rural sector, with more than 90 per cent of peasant households using such a system.

The key element in this system is the production contract. Signed by the peasant household and the collective administrative unit, the contract clearly defines the responsibilities of the authorities and the individual household. It places a specific piece of 'responsibility' land at the disposal of the contracting household. The household can utilise this land to its best advantage, although in return for the use of the land it must fulfil its contracted quotas for both state procurement and collective retentions. It must also provide for the subsistence of the contracting family. Once these demands are met, the household is entitled to any output from that land and may sell it to the state at an above-quota price, or privately at local markets.

In practice, this often means that the individual household is under a contractual obligation to produce a certain amount of grain or other specified crop (e.g. cotton) to be sold to the state at a low price. Grain will also be produced for collective retention and personal use. Once these demands have been met, the household is more likely to devote its land, labour and capital resources to cultivate cash crops and develop sideline occupations which command better prices.

The major advantage of the production responsibility system is that it implements the principle of 'to each according to his or her work'. The peasant can see that hard work means in theory more reward, and enthusiasm for work is encouraged. This is a clear attempt to install material incentives within the rural economy following the egalitarian policies which had previously prevailed. Rises in official prices for rural outputs of around 25 per cent also strengthened incentives.

Nevertheless, the disadvantages of this system – potential and actual – should not be overlooked. The most immediate set of problems are those associated with the contract itself. There are difficulties because contracts are not honoured, sometimes by the peasants failing to fulfil quotas, sometimes by the authorities who arbitrarily raise quotas and increase collective requirements. Another concern is the further creation of surplus labour in a rural economy where unemployment and under-employment are widespread. As much as 30 per cent of the rural labour force is considered to be surplus, and it is conceded in China that responsibility systems add to this figure.

Perhaps the area of greatest concern is the emerging relationship between the individual peasant household and the village unit. While not strictly owner-occupiers, peasants have certainly developed a system of family farming. There can be no doubt that after the Maoist excesses the majority of peasants have seized the chance to work for themselves, relying on their own skills without having to worry about the work attitude of others. Recent announcements even suggest that peasants will be permitted to buy and sell the rights to use land, so encouraging the emergence of a smaller number of 'successful' farm enterprises. This shift towards farming has weakened the collective units in a number of ways. Some commentators suggest that the collective can no longer effectively maintain and develop the agricultural resource base. Important tasks such as irrigation and drainage work, agricultural machinery maintenance and repair, and disease and pest control are being neglected.

This neglect has arisen through a combination of factors. Too many rural administrators are more concerned with making money themselves than devoting the necessary time and energy to collective work. Furthermore, many collectives have found it increasingly difficult to accumulate the funds and labour needed to maintain and expand rural production. The crucial difficulty is that peasant households themselves now have responsibility for agricultural production, including the purchasing of producer goods, the maintenance of irrigation ditches and so forth, previously managed by the collective.

The emergence of family farming not only weakens the collectives' ability to perform key economic tasks; it has also lessened the extent of social control in the countryside. Signs of this weakening social control are evident in numerous reports of misuse of collective land, illegal building of houses by peasants on good cultivable land and, formerly, the illegal transfer and leasing of land. Collective welfare provision has also suffered from the weakening of collective control of rural resources.

These disadvantages are important. For some observers, they contradict all that socialism in rural China has stood for. Moreover, although family farming, under the auspices of the household responsibility system, has worked well to promote immediate short-term growth as reflected in increases in per-unit yields, it is inevitable that in the long term natural constraints will limit the extent to which enthusiasm for production can be a substitute for an improved agricultural resource base and advanced technical and management methods. It remains to be seen whether a system based upon family farming can effectively make the necessary investments.

Specialisation in the rural economy

Alongside the development of family farming, there has been growth in product specialisation in the Chinese countryside. This represents a distinct change from the Maoist model of rural development where units were encouraged to be self-sufficient. In this growth, 'specialised households' have become an important force. As suggested earlier, the household responsibility system has encouraged peasants with special production skills to develop them, sometimes in conjunction with other farming activities, sometimes full-time. Growing numbers of peasant households throughout China are now identified as specialists, with a significant amount of a household's agricultural output and income derived from a specialised occupation. By the end of 1984, 13.6 per cent (about 23.4 million) of all peasant households were said to be specialists.

Specialist households are important. It is argued that they can more effectively utilise local resources to satisfy local and national needs. The post-Maoist period has seen a shift away from an emphasis upon poorly rewarded grain production towards a policy of diversification in agricultural production. Specialist farm producers have been to the fore in developing diversified farm undertakings. Similarly, such producers have been prominent in taking advantage of the post-Maoist rural economic environment to find employment in handicrafts, rural industry, construction and commerce. Growing pressure on land resources requires increasing employment created in non-agricultural activities. Specialised households play an important role in this development by creating non-farm employment.

By marrying local peasant skills to resources, it is argued that it is possible to maximise returns on investment. Numerous case studies indicate that per-unit yields achieved by specialised households are significantly higher than those by non-specialists. Individual income from specialised production is also often higher than that to be gained in local factories (traditionally the best-paid rural occupation) and much higher than that to be found in non-specialist farming. Unfortunately, such success is at risk of being a 'state-sponsored' divergence of richer and poorer households. In some places this visible inequality has resulted in bitter local tensions and conflict.

Organisations of specialised households also exist. These can range from small-scale co-operation among twenty households, through village organisations, to county-based units of a thousand specialised households or more.

Beyond the county level, larger areas of more general specialisation are identified as 'commodity production bases'. These bases are essentially areas (comprised of as many as fifteen counties or more) whose physical characteristics favour concentration upon a particular crop (e.g. cotton, tobacco, tea), the production of which is already well-established locally. The state builds on this foundation (for example, through appropriate investment in processing equipment and commercial channels) in order to maximise production of that product. In this way, the state can not only satisfy growing domestic demands for agricultural produce, but also maintain a greater degree of control over agricultural production. Cotton in southern Hebei is a case in point.

Nevertheless, just as local disputes between specialist and non-specialist households reflect growing disparities of income through product specialisation, similar conflicts can be seen between those areas which are able to specialise profitably (often with state support) and those which are denied the opportunity to do so. There is evidence for this in sugar, cotton, edible oil and other products.

Commodity production and commercial channels

Since 1978, the state has actively encouraged individual peasant households to increase production by giving them more control not only over rural production but also the disposal of their produce, involving them in commodity production and trade. Specialised production is an extension of such a policy. This is an important new direction for the rural economy.

The success of this has been founded on the sharp improvement shown in the production of basic foodstuffs, especially grain, and that of important agricultural raw materials in the late 1970s. While it is admitted that some backward areas (where basic food and clothing requirements have yet to be met) still exist, production of agricultural staples is considered sufficiently improved in the current phase to allow the expansion of cash crop production and sideline occupations. Thus, the peasants have the opportunity to improve their incomes through commodity sales in the market-place.

There are, of course, constraints upon such a policy, notably the need to produce sufficient quantities of agricultural staples (grain in particular) to meet the ever-increasing domestic demand. There is also the need for an efficient commercial system through which peasants can sell surplus produce, and also purchase necessary producer goods such as fertiliser, fuel, machine parts and pesticides. It has become increasingly evident since 1978 that the rural commercial system is antiquated, bureaucratic, unwieldy and unable to handle the enlivened rural sector. Two groups of problems can be identified in rural commerce: organisational and infra- structural. Bureaucratic mismanagement in the current phase is mainly attributed to the negative influences of the Maoists. The actual number of retail outlets in 1979, for example, was only 200,000, compared to 1.01 million in 1957.

There is a grave shortage of qualified commercial personnel. Bureaucrats are now expected to respond to market conditions. Not surprisingly, this causes its own problems. The increasingly complex nature of rural production in turn places great strains on the commercial channels. Infrastructural problems also abound. Rural transport links are weak. Roads, the cornerstone of the rural transport network, are often narrow, poor in quality and unevenly distributed. In 1984, it was estimated that 27.8 per cent of all production brigades were not physically accessible by truck.

The means of transport are similarly backward. Total carrying capacity is too small, services are few in number and seldom prompt, haulage fees are high and vehicle efficiency is low. Almost half of China's total cargo capacity in the rural sector is in the form of tractor transport, a slow and expensive form. The use of tractors in transport also limits their availability for farming; but tractors are more widely available to the rural people than trucks. Finally, storage facilities are poor. In many rural areas the state quotas for grain often exceed the local storage capacity, leaving stockpiles of grain exposed to the elements and vermin. Losses are 'enormous', according to a Chinese critic. The same is true for other kinds of agricultural produce.

Reform of the official commercial system has been slow, confused and patchy. While increase of specialist households engaged in rural commerce is widespread (most rural retail businesses are now individually owned) the scope of individual activity remains extremely limited. The state remains very reluctant to loosen its grip on the virtual monopoly it has over rural wholesaling and agricultural prices, and for the time being trade over large areas is difficult, and perceived as very risky, for individuals.

Real improvement awaits both expensive investment in road, storage and so forth, and also the state's willingness to accept further loss of control of rural commerce.

THE GRAIN SYSTEM

Under the Maoists, grain production was regulated through the rural planning system, and trade in grain was a state monopoly. Consumption was almost always rationed in some sense. At that time, the grain system was one arm of official control in the community − not only economic control (through low official purchase prices combined with the obligation to produce), but also social control (where the village's obligation to produce grain for state purchase often dominated the year's work). Inputs to the grain system (notably artificial fertilisers) increased considerably,

Table 6.3 Agricultural outputs for various years, 1949–85

	Grain (millions of tons)	Cotton (millions of tons)	Sugar crops (millions of tons)	Oilseeds (millions of tons)	Pork beef and mutton (millions of tons)	Total population (millions)
1952	164.9	1.3	7.6	4.1	3.4	575
1965	194.5	2.1	15.4	3.6	5.5	725
1978	304.8	2.2	23.8	5.2	8.6	963
1984	407.3	6.3	47.7	11.9	15.4	1,035
1987	405.0	4.2	55.5	15.3	19.9	1,081

Sources: Statistical Yearbook of China (annual); Zhongguo Foingji Nianjian 1988 (for 1987).
Note: Figures are given for 1952, by which time the country was stabilised under the People's Government; for 1965, before the Cultural Revolution; for 1978, the year in which the new post-Mao policies were formulated, 1984 the peak year so far in grain production, and 1987.
Most of the outputs shown doubled or nearly doubled between 1952 and 1978 (while the population rose by 67 per cent), but not that of oilseeds. Since 1978, increases of output have been very much more rapid – even of grain, the main Maoist preoccupation.

but nevertheless before 1978 grain output did not rise above 290 million tons (Table 6.3).

After 1978, though, grain output rose to levels consistently above 300 millions tons, and in 1984 to a record 407 million tons (Table 6.4). The increases in output were stimulated by price increases. They were also based on sound agronomic change, leading to an increase in unit area yield of 43 per cent in 1984 over 1978 (in the same period grain acreage fell slightly). In grain, which the Maoists made the heart of their policy in the countryside, Maoist achievements were strictly limited, and have since been clearly overtaken.

The grain system has been radically changed since 1978. Local subsistence is now the responsibility of individual households farming their own land contracted in various ways from local collective organisations. Grain purchased by the state is now produced, increasingly, not by villages with quotas to fulfil, but by specialist households contracted to produce it. Grain-growing peasants are guaranteed a state minimum price for any additional grain delivered, and according to its needs the state may offer a higher price for additional grain. Alternatively, since the state's monopoly in grain dealing has now been dissolved, the peasant can sell above-contract grain on the open market – to breweries, distilleries or bakeries, or for animal or poultry feed.

Inputs of fertiliser and pesticide have continued to increase sharply; but it is difficult to refute the belief of the authorities that what has changed most has been the peasants' motivation. Basic prices have risen, but of course prices of grain sold on the open market have risen much more. In good years, like 1984, China had as much grain as it could handle. Interestingly, the amount of tractor-ploughed land has fallen since 1978, by some 20 per cent, partly no doubt because tractors are used as commercial

Table 6.4 Rural progress since 1978

Year	Sown areas (millions of ha)		Crop yields (kg/ha)		Rural markets		Rural inputs	
	Grain crops	Industrial crops	Grain crops	Cotton	Numbers of markets (thousands)	Value of trade (100 million RMB)	Rural electricity (100 million kWh)	Chemical fertiliser (millions of tons)
1978	150	14	2,528	443	33	125	253	9
1981	145	18	2,828	570	40	253	370	13
1984	144	19	3,608	904	50	390	464	17
1987	111	21	3,630	870	59	811	659	20

Source: Statistical Yearbook of China (annual); Zhongguo Fongji Nianjian, 1988 (for 1987).
Note: These figures are a selection from those which are available, to illustrate the present phase of rapid growth in the rural economies of most parts of China. Sown areas (which include double-cropping where that is practised) have fallen for grain crops but risen for industrial crops such as cotton and sugar. Crop yields have increased dramatically. Figures for markets represent rapid increases in rural trade. Corresponding levels of increase are also recorded for rural inputs, illustrated by electricity and artificial fertiliser.

vehicles, whilst animals pull the plough; partly because tractors are expensive to use, and partly because family fields are often too small.

The peak year for grain output in the period 1979–88 was 1984, and since then output has been weakened by the continuing conflict between high costs (of inputs) and prices which are still low. This has led to farmers switching to alternative crops or other kinds of rural production, and in more recent years by migration to towns and cities. Grain is the heart of a general weakening of growth in agricultural output, the causes of which are increasingly being discussed.

DIVERSIFICATION: REGIONAL AND LOCAL REALITIES

Diversification, re-emphasised since 1978, is of course implicit in the state's commitment to give peasants more freedom over rural production, and its re-emphasis has had a number of important effects within the rural economy. Fields inappropriately claimed for the cultivation of grain at the height of the 'take grain as the key link' period began to be reclaimed for their original use.

While this move had important local ecological benefits, the tendency to reduce grain farming is due more to peasant inclination to grow cash crops rather than less well-rewarded grain crops. Income gains rather than any rational restructuring of the farming economy is the main motivation.

Nevertheless, it must be noted that the freedom to diversify production to try for income gains, and the ability to do so, do not always go hand-in-hand. Cultivable land may be poor, the climate inhospitable, and capital resources to improve the arable unavailable. Not only might the environment be incapable of producing a variety of cash crops, it may also be that what resources are available have to be concentrated on subsistence production. Furthermore, in remote areas, while diversification may be possible, access to markets may be very limited. These are some of the factors which lead towards marked spatial disparity of development and prosperity in the present phase.

Obvious differences are found even within local areas. Ganyu county in northern Jiangsu, for example, is a rapidly developing agricultural area, although still significantly poorer than its neighbours to the south. Ganyu has an undulating landscape. In the valleys, farm undertakings are notably diversified: cash crops include mushrooms, beans, tomatoes, asparagus, peppermint and mulberry bushes; goats, rabbits and pigs are bred; fish-farming is widespread. These activities indicate accessibility to sizeable local markets, and an unusual level of technical capacity. The reverse is true of counties in the upland, where the emphasis remains upon subsistence grain production. Income differences are marked and growing.

RURAL INDUSTRY: AN ENGINE FOR RURAL GROWTH

Current policy concerning rural industries has emphasised the development of village and township industrial enterprises. These industries are a significant source of both peasant income and employment opportunities, and also collective accumulation funds. This latter point is important. Between 1979 and 1982, while the state invested 11 billion yuan in agriculture, animal husbandry and water conservation projects, contributions from rural industrial enterprise to farmland construction and the purchase of farm machinery amounted to 8 billion yuan.

Areas of dense population

The impact of developed local industry can be seen in Wuxi county, Jiangsu province (Figure 6.4). Wuxi has a long history of rural industrialisation (see also Case Study 8.1 on p.220). Recently, emphasis has been placed on the development of agricultural-machinery repair and spare-parts factories, and grain- and silage-processing enterprises. The county benefits from rural industry in a number of ways: first, it employs a growing amount of rural labour; second, it provides – in situ – much-needed goods and services to the rural economy; third, it provides up to 90 per cent of the county's investment funds for basic field construction and agricultural mechanisation; fourth, it can provide a stable demand for raw materials, encouraging peasants to invest in production; and, finally, funds from rural industry can also finance subsidies to farmers in order to reduce the income differences between farming and industrial occupations. This is one way in which grain-specialist households may be subsidised.

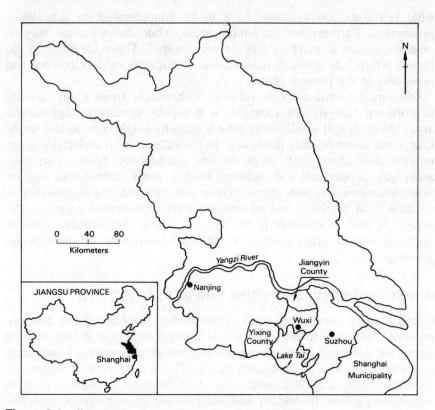

Figure 6.4 Jiangsu province and the location of Wuxi

Wuxi is exceptional. The successful development of rural industry requires the availability of local resources – both raw materials and capital; good transport links and commercial channels to large local markets; and a sophisticated rural production system. All these exist in Wuxi. Furthermore, Wuxi county benefits by being administratively controlled by Wuxi city, linking the rural enterprises with the more developed, large-scale city industry. A 'putting-out' relationship is developing between these rural and urban industries, firmly establishing the role of the rural concerns. These conditions are not common in China, but the state would like to see them extended.

The development of rural industry has also encountered difficulties. Peasants prefer to leave farming for industrial jobs, which has adversely affected farming production, especially of grain; and industrial development has often been irrational, with much duplication of production and increasing local competition. Rural industry also competes with state industry for markets and raw materials. Its development in some areas often serves to accentuate patterns of rural income differentiation. Yet the contribution of rural industry to local economies cannot be denied.

The remoter countryside

The Communist Party in general, and the Maoists is particular, have laid much stress on the achievements of the People's Government in rural development in remote areas of China: construction of farmland from waste in the northeast, the desert frontier zone, and the mountains everywhere; extension of agriculture in the western half of the country; rural electrification, the construction of dams for irrigation, and the improvement of communications. Some of their achievements, particularly the extension of arable land-uses on the erosion-prone loess plateau of the north, have turned out to be environmentally destructive; some, particularly in the northeast, unnecessarily extravagant of land; others, particularly some dam projects, misconceived.

But the more important weakness of these highly publicised programmes has been their limited scale. Of 2,046 counties, about half may be reckoned mountainous or otherwise isolated counties, and of these the great majority are still seriously short of transport (vehicles as well as roads); have serious problems of land resource or land management or both; and have low standards of living which are very difficult to raise by any local effort. Maoist policy, apart from a few highly publicised models, had only feeble effects in most of the backwoods counties and villages.

The same has to be said of present policies. The community achievements based upon urban demand outlined in the last section (and Case Study 8.1 on p. 220) are absent in places where urban demand is limited and ineffectual, or where isolation locks together town and countryside in a low-output, low-growth, low-achievement economy.

POLICY, SPATIAL INEQUALITY AND ENVIRONMENT

The standpoint of the present regime is that in due course, by emulation and the spread of both better techniques and stronger demand, prosperity will diffuse through the rural areas to the backwoods. The weakness of this reasoning is the same as those of the Maoist campaigns for physical renewal: its inadequacies of scale. In relatively favoured areas on the edges of the mountain masses, particularly in eastern China, prosperity has already begun to diffuse in this way; but in the great bulk of backwoods counties in such provinces as Sichuan, Yunnan and Shaanxi it cannot do so in the foreseeable future; and in the western half of China (Xinjiang, Tibet, Qinghai) any such development will have to be locally based and can also affect only a few favoured areas.

The Chinese forests are a special problem. In spite of purposeful afforestation programmes in the 1950s, the total afforested area has declined since 1949, due to campaigns for more arable, for rural industrialisation (which uses great quantities of firewood) and (recently) widespread rural building, which results from increased prosperity – as well as increased urban demand. In recent years, timber imports have been increased in an attempt to satisfy demand in the cities. Meanwhile, the rural communities are making increased demands: timber for building, and firewood for cooking, rural industry, and heating stoves in winter. Accessible forests everywhere are under sharp pressure from local communities and even from pirate enterprises which cut state timber illegally in remote areas (see Chapter ten).

Reafforestation programmes have been stepped up but are still inadequate. Difficult terrain and lack of transport are the main protectors of the remaining forests, for instance in the mountain peripheries of the Tibetan massif, in Yunnan and Sichuan, and in the high mountain ranges of the northeast. The Chinese forests, particularly in the south, are exceptionally rich in species – not only trees and other plants, but animals, birds and bugs which find special ecological niches in the rich flora. Such complex ecologies are easily destroyed and if seriously disturbed can seldom recover completely.

EMERGING PROBLEMS

What has become clear since 1978 is that the Chinese rural economy has demonstrated a real improvement in production output and is displaying a new vitality. Peasants are enjoying real income gains. However, while the rural sector as a whole has shown improvements, some areas and groups in rural society are significantly better off than others and are tending to become more so.

Development today can be seen as the opportunity to diversify beyond grain farming, and beyond the arable, as local agricultural resource bases and state production quotas permit. Patterns of spatial differentiation are little changed now from those displayed in pre-Liberation China. The essentials are the difference between the production conditions of north and south China; the continued development of the coastal provinces faster than the interior; the inability of mountain communities effectively to utilise their environments to generate wealth and achieve the levels of prosperity attained on the plains; and the continued neglect of the 'peripheries', in particular the provinces of Xinjiang, Qinghai, Tibet, Gansu, Yunnan, Ningxia and Inner Mongolia, where the bulk of China's ethnic minority population is located.

Again, current policy represents a move towards a more sophisticated, complex and 'modern' market economy, encouraging as it does commodity production and trade. Yet, for those peasants who cannot develop cash crops and sideline occupations because of the demands made by grain farming, those physically isolated from markets and commercial channels, and those without rural industrial opportunities, the potential gains of commodity production remain elusive. For these peasants, current policy does little except compound natural and historical distinctions and consolidate the traditional patterns of differentiation. The failure of the state to change these patterns may not be unexpected given the scarcity of funds.

Difficulties in rural production systems

Tensions and conflicts are apparent in the implementation of the new rural production systems. This has been most noticeable in the allocation of responsibility land. No single method of land allocation has been adopted. Some localities assign land to families on a per capita basis; others, according to families' labour-forces; a third way is a combination of the first two; and a fourth method is purely on the basis of ability, with the grain ration of those households which receive little land (perhaps because they are working in rural industry) guaranteed through collective retention.

The first three methods give rise to many contradictions. Obviously any locality will contain a wide range of production skills and capabilities. Thus, a pro rata distribution will fail to match land requirements with land allocation. Pro rata distribution can also present the state with ideological difficulties. Land allocation on a per capita basis, for example, can result in a household's having land without the labour necessary to till it.

In this situation the household could have either illegally rented out the land or hired labour to till it on their behalf. Despite its *de facto* emergence, until recently the state was unwilling for such 'landlord–tenant' and wage–labour relationships to develop. But the March 1988 decision to allow the sale of land-use rights effectively permits such relationships. Per capita distribution also serves to aid the viability of large households, something which the state is trying to counter in population policy (see Chapter five).

It would seem most logical to allocate land according to ability, but this is often strongly resisted because of long-standing peasant regard for the land as means of subsistence. Few families would relish the prospect of relying on collective distribution of grain; memories of grain shortages in many rural areas are still too fresh. Ironically, pro rata allocation may serve to inhibit rural development. Prior to allocation, land is graded according to its quality (soil quality, irrigation conditions, accessibility from the village, and so forth) with distribution ensuring that each household receives an appropriate amount of a variety of graded pieces.

The process of proportionate land division leads to excessive dispersal of farmland holdings, which is unfavourable for intensive or mechanised production. Where rural production is more sophisticated and specialisation is more prevalent, this problem can be overcome by land consolidation through transfer, but this has its own problems.

A rural exodus?

Through strict social control, the state has usually been able to control the movement of people within China. In particular, it was able to prevent the migration of the rural population to the urban areas in the kind of numbers which have burdened other Third World cities (see Case Study 5.1 on p. 126 and Chapter eight). But, as has been said, labour is surplus in rural China to the extent of up to 30 per cent, or locally even higher.

Currently, there are signs that increasing numbers of peasants are moving to urban areas (including small towns) in search of better opportunities and lifestyles. Some of these peasants, with marketable skills, may receive permission to live in the town or city, but for many such residence is illegal. Despite this, at least 8 million people of work-age are entering the cities annually. The consequence could be the overcrowded cities all too familiar throughout the Third World.

There is another aspect of current rural development which may be seen as a form of rural exodus. This is the move towards employment of rural people in non-farm occupations, in particular rural industry. The importance of rural industry to the rural sector has been well-established. However, in some areas the growth of rural industry is such that peasants are leaving the land in significant numbers, creating labour shortages in periods of labour-intensive activity such as harvesting.

Furthermore, with some peasants leaving farming it is becoming clear that in some areas land is increasingly concentrated in the hands of fewer families. Concentration of land is an ideal breeding-ground for wage–labour and 'landlord–tenant' relationships. Contracts for 'responsibility' land are now often for periods of 15 years, or even more, which must have a parallel effect. However, in the pursuit of rural prosperity, tensions between some socialist ideals and some economic goals are inevitable.

A FRESH START

In the long perspective of the history of the Chinese countryside the present phase does indeed represent a fresh start. The countryside is tranquil and busy. The landlords are gone; the land is still formally the property of local collective units, normally the villages. The Maoist obsession with egalitarianism is also gone; trade and production for trade are encouraged. Socialism now seeks to promote the welfare of working people in material terms. Some of the natural pests of farm countrysides –

plant, animal and human diseases – can be effectively treated, thanks to scientific methods. Some of the drudgery of work on the land can be given up, thanks to farm machinery, diesel engines and electricity.Village houses can be rebuilt, sometimes with electricity and running water. In the best areas of eastern China, particularly in the south and near the cities, the incomes and amenities of rural life rival those of the increasingly congested cities. In the good areas, the present has the qualities not only of a fresh start, but of a golden age.

Of course, all is not perfect. At many points in this chapter details have been given of problems of various kinds which arise in the new management systems. It has been shown that in many peripheral and mountain localities the new prosperity has been slow to penetrate. Various kinds of problems and corruption beset the organisation of family farming, particularly in the difficult field of land-holding. Local industrial enterprises may be little more than sweat shops. Village heads may be aggressive and self-seeking. Arbitrary and corrupt officials have not disappeared. Taxation, hard work and monotonous food all remain. Many rural people resent the one-child family policy, particularly those who could well afford a larger family. Prosperity is still conditional upon development; it is not universal and demands intelligent effort.

Official policy since 1984 proposes new, radical changes in rural China, including the shedding of up to 30 per cent of labour from the land and its redeployment in rural industry or (as some argue) in the cities. This would involve at least 100 million workers. Change of this kind would at last reverse the long-term intensification of Chinese agriculture through increased labour inputs, and would look towards a commercialised countryside very different from the self-sufficiency of the past. Changes on these lines would be no less traumatic than those in the past decade, or those experienced under the Maoists.

FURTHER READING

Barker, R., Sinha, R. and Rose, B. (eds) (1982) *The Chinese Agricultural Economy*, Boulder: Westview Press.
  Aspects of agricultural policy and performance reviewed by leading Western specialists.
The China Quarterly (1988) 116: 'Food and agriculture in China during the Post-Mao era'.
  A special issue of the journal. Includes articles by leading specialists analysing policy and production, food consumption, agricultural planning and organisation, surplus labour and agricultural technology.
Endicott, S. (1988) *Red Earth: Revolution in a Sichuan Village*, London: I.B.Tauris.
  One of very few village studies of China in which the author has been able to return to the same place over a number of decades to give the history of political and economic events as experienced at local level.
Hinton, W. (1968) *Fanshen*, New York: Vintage Books (and other editions).

Classic eye-witness account of the changes in a village in north China during and after the 1950s land reform. The author is an agriculture specialist who lived in China.

Hinton, W. (1985) *Shenfan*, London: Picador.
Sequel to *Fanshen*.

Leeming, F. (1985) *Rural China Today*, London: Longman.
Contrasts the developing family farm system with the collective system under the Maoists. Individual treatment of major regions.

Luo Hanxian (1985) *Economic Changes in Rural China*, Beijing: New World Press.
Reviews the rationales behind the individual enterprise revolution in rural China, with an abundance of case studies. State Statistical Bureau.

Mosher, S. (1983) *Broken Earth: The Rural Chinese*, New York: Free Press.
Account by an American anthropologist allowed to live and research in south China. It is a hard-hitting, journalistic indictment of much of post-1949 policy; compare his account with Hinton's. He was expelled from the country after a dispute over his criticisms of the population-control policy and use of abortion.

Saith, A. (ed.) (1987) *The Re-Emergence of the Chinese Peasantry*, London: Croom Helm/Routledge.
Essays which explore the reasons for and consequences of the return to family household farming and enterprise.

Case Study 6.1

NANJING: RURAL–URBAN RELATIONSHIPS

Simon Powell and Frank Leeming

Belts of comparative affluence surround most Chinese cities, due to the opportunities presented by the urban market. Nanjing is a good example (Figure 6.5). Situated on a plain in southwest Jiangsu province, Nanjing is surrounded by a terrain of low hills and rivers, most notably the Chang Jiang River (Yangzi). The urban fringe of Nanjing consists of five counties administered by the city – Jiangpu, Jiangning, Liuhe, Lishui and Gaochun – with 3.3 million mu (equivalent to one fifteenth of a hectare) of arable land. The periphery also contains a diversity of mineral wealth, including deposits of iron, gypsum, lead, zinc and manganese, as well as limestone and sandstone. Agriculture and rural industry are both well-developed.

Vegetable production is concentrated in the areas closest to the city, with relatively easy access to large urban markets. This, combined with the high unit-value of vegetables, enables producers to enjoy comparative prosperity. Beyond the predominantly vegetable-growing areas, agriculture is highly diversified and specialised. Grain remains an important crop,

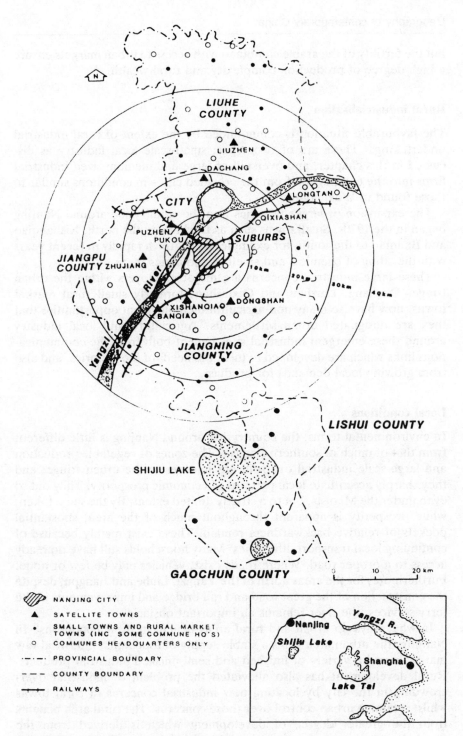

Figure 6.5 Nanjing and its urban fringe

LIUHE
COUNTY

LIUZHEN

DACHANG

CITY

LONGTAN

QIXIASHAN

PUZHEN
PUKOU

SUBURBS

JIANGPU
COUNTY   ZHUJIANG

River

10km

20km   30km   40km

DONGSHAN

XISHANQIAO
BANQIAO

JIANGNING
COUNTY

Yangzi

LISHUI COUNTY

SHIJIU LAKE

GAOCHUN COUNTY

0   10   20 km

NANJING CITY
SATELLITE TOWNS
SMALL TOWNS AND RURAL MARKET
TOWNS (INC SOME COMMUNE HQ'S)
COMMUNES HEADQUARTERS ONLY
PROVINCIAL BOUNDARY
COUNTY BOUNDARY
RAILWAYS

Nanjing   Yangzi R.
Shijiu Lake

Shanghai

Lake Tai

but the fertility of the arable combined with access to urban markets ensure a high degree of production complexity and rural wealth.

Rural industrialisation

The favourable situation is compounded by the extent of rural industrial undertakings. These are of two types: small-scale local industry as discussed in this chapter, and overspill of large and medium-sized industrial firms into the countryside from the crowded city – in conditions similar to those found in Wuxi county.

The expansion of larger factories into the countryside around Nanjing began in the 1970s. Small towns such as Dachang to the north, Xishanqiao and Banqiao to the south, for example, have grown rapidly in recent years with the siting of chemical and smelting industries there.

These large industrial concerns are major employers within the urban fringe. Dachang, Liuzhen and Qixiashan, formerly small rural market towns, now have so many non-agricultural employment opportunities that they are designated urban settlements. Agriculture and local industry around these emergent industrial sites benefit both from the communications links which are developed to meet the needs of the factories, and also from growing local demands for produce.

Local conditions

In environmental terms, the countryside around Nanjing is little different from that of much of southern Jiangsu. The zones of vegetable production and large-scale industrial overspill are features of the urban fringe; and they sharply accentuate local patterns of economic prosperity. They did so even under the Maoists, but to a strictly limited extent. By the same token, while prosperity is apparent throughout much of the area, substantial pockets of relative backwardness remain. These exist mainly because of continuing local transport difficulties. Many households still have no ready access to a proper road. Where roads exist, vehicles may be few or none. Furthermore, for the areas north of the Yangzi, Liuhe and Jiangpu, despite the construction of the great road and rail bridge and improvements in the ferry services, the river remains an important obstacle.

In what ways do urban and rural areas benefit in these conditions? In Nanjing, the urban area gains a stable supply of foodstuffs, industrial raw materials and a variety of finished and semi-finished industrial products. Rural development has also alleviated the problem of industrial overcrowding in the city by locating new industrial concerns in rural towns whilst retaining urban control over those concerns. The rural area benefits from the greater degree of development which is derived from the proximity of a large urban market and industrial overspill. However, this development is uneven. In the countryside, consumer and producer goods remain difficult to obtain, and standards of hygiene, education and recreation facilities are still comparatively low. Important tensions remain within rural–urban relationships.

The conditions which have been outlined for Nanjing apply, in broad terms, to all Chinese cities. They apply also in a range of localities where urbanisation and industrialisation are strong, such as that lying between Nanjing and Shanghai, and that centred on Beijing, Tianjin and Shenyang. Policy is pressing on with moves to integrate cities more closely with neighbouring rural areas, with the deliberate intention of generating constructive urban–rural contacts.

Case Study 6.2

FROM COLLECTIVES TO FAMILY FARM MANAGEMENT

Frank Leeming and Simon Powell

Collective management: the People's Communes

In the People's Commune system, practically all arable land was collectively used, and there was no individual ownership of land or means of production. People belonged to work teams (usually a sub-division of the natural village) and these, in turn, were subordinate to production brigades. These brigades were themselves a division of the commune. The commune acted as production co-ordinator between the government and villagers, using the three-tier system as both a structure for organising production and an administrative and political hierarchy.

Figure 6.6 illustrates, in simplified form, the workings of the rural economy in the communes before 1979. Working upwards from the bottom of the diagram, the inputs ('means of production') needed in farming or local industry were supplied by the collective to the different sub-units of the commune. These were used on the arable land (next rectangle) by the farming workforce, and in village industry; in return, they provided the commune with grain and investment funds. Also from the collective's output came the grain to pay agricultural tax to the state, and the commune also collected a share of the output to pay for future investment and welfare spending.

The people earned work points instead of wages: each day's effort would

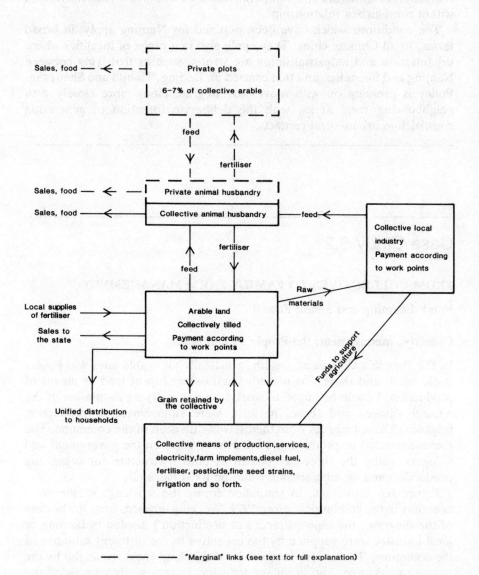

Figure 6.6 Diagram of the structure of rural economy before 1979

be rewarded by work points up to a maximum of ten (though women were normally only allowed to earn a lower maximum). A work point was not worth any fixed sum. Rather it was like a dividend; at the end of the season when the harvest was complete each work point would be worth a proportion of that output, depending on how many points had been earned overall. Each person would then be allocated a share of the total available for distribution depending on how many points had been earned by that individual.

For example, if the output available for payment to members of the collective was 100 tonnes of rice, and during the season everyone had earned a total of 100,000 work points, each point would be worth 0.001 tonnes. So a person who had earned 500 work points during that season would be entitled to 0.5 tonnes of rice.

Some produce was sold to the state. These sales financed the purchase of producer goods and the distribution of some cash for work points. The state might also supply quantities of fertiliser – perennially in short supply – to the collective. Often, such supplies were related to the amount of grain sold to the state.

Animal husbandry (especially pig-raising, fish ponds and poultry) and other 'sidelines' in the collective (next rectangle in diagram) were also rewarded using the work-point system, and would earn cash from the state or nearby cities for the collective. The 'marginal links' indicate that part of the rural economy under the Maoists which was theoretically 'private' (top rectangle). Between 5 and 7 per cent of the collective land was normally reserved as private plots.

While the Maoists did not actually prevent the use of such plots, many local cadres blindly followed the lead of Maoist models such as Dazhai (a brigade in Shanxi province) where private plots were incorporated into the collective arable. This was significant, as peasants had traditionally grown cash-crops and developed sideline occupations such as animal husbandry on this land. Such production was an important source both of foodstuffs and household income.

Family management in the post-1979 reforms

Figure 6.7 simplifies the workings of the post-1979 rural economy. The collective arable was divided amongst individual households through production contracts, which specify the household output quotas set by the collective and state, and also contributions to collective funds. Any output which a family produces over and above this is used for its own consumption, and any extra may be sold to the state or privately. The intention of this change is to provide a much greater incentive for families to produce more than before, by virtue of there being a more direct link between their effort and reward.

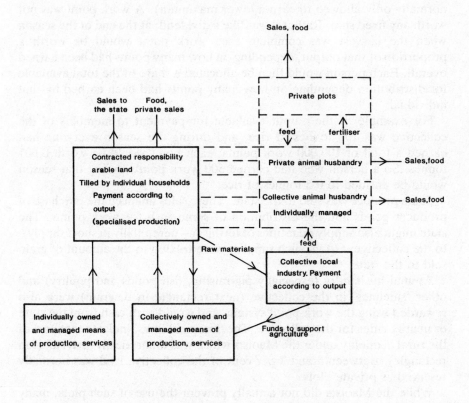

Figure 6.7 Diagram of the structure of rural economy after 1979

Workers in collectively run local industry are also paid according to output. The collective continues to supply some of the means of production, either as part of the contract or at a price. Households may contract jobs to individuals who specialise in agricultural services, such as ploughing. The collective also continues agricultural investment. This is made possible by household contributions to the collective funds and – increasingly important – by contributions from local industry under the control of the local authority.

The 'blurred links' indicate a new development in the rural economy. In 1978 the status of private plots was reinforced and increased (up to 15 per cent of the collective). However, with the development of family contract farming, private plots have rapidly become merged into the households' contracted resources, losing any separate identity. In many areas, the same is true of collective animal husbandry. While some units have contracted out animal husbandry to individual households in the same way as collective arable, other units have simply distributed previously collective animals to individuals.

Chapter seven

Industry, energy and transport: problems and policies

Maurice Howard

The intention here is to look at the central concerns of Chinese economic policy outside the agricultural sector, industry in particular, and what are arguably the two sectors which most constrain the development process – energy and transport capacity. China's course of development has placed particular burdens on these last two sectors, which have contributed to shaping the nature of China's industrialisation. The geographical dimension, as we shall see, has been particularly important. Energy and transport will be analysed later, after examining the nature and location of industry.

INDUSTRY

A useful idea of the structure of industry can be had by looking at comparisons with countries of similar size or level of development. In these two respects, China and India are quite close. Both have a large proportion (around 70 per cent) of their employment in agriculture. In China, industry employs around 18 per cent of the labour-force, depending on definition, a high proportion for its level of income.

Also worth noting is far lower employment in services (mainly commerce), and a higher proportion in manufacturing and mining, than India. This discrepancy is even greater when compared with less agricultural developing countries such as Indonesia, Pakistan, or Korea.

The share of agriculture in China's national income has tended to decline in recent decades (from 68 per cent in 1949 to 35 per cent in 1978). The economic reforms have reversed this, at least temporarily, and agriculture's share of output, at over 40 per cent, is, for the first time since 1970, on a par with industry.

Employment in heavy industry (which provides equipment and supplies such as steel, chemicals, and machine-tools to other industries) is quite high as a proportion of the total, nearing the levels in industrialising countries such as South Korea and Taiwan. On the other hand, light industries (producing consumer goods) and services of all kinds employ a noticeably smaller percentage of workers, even when compared with the USSR.

These proportions reflect generally consistent priorities which remained the orthodoxy until reforms began in 1979. Heavy industry was given priority over light, manufacturing over sevices, and 'productive' over 'non-productive'. Following Soviet practice, this was seen as a way of accelerating development. Inevitably, these priorities have affected industry geographically.

The distribution of industry in China is very uneven. The coastal provinces, forming 14 per cent of the area of the country, account for over 60 per cent of industrial production. The three adjacent provinces of Jiangsu, Shanghai and Zhejiang alone account for well over a quarter of the national output. A further concentration appears in the three north-eastern provinces (Liaoning, Jilin and Heilongjiang), which contribute nearly one-eighth of output.

These concentrations of industry are not due merely to large populations in these areas (the two groups of provinces represent only 11 per cent and 9 per cent respectively of the total population). So it is worth considering in some detail the reasons for their existence, before going on to look at the way policy on the location of industry has changed since 1978.

The industrial legacy

Before 1949 'modern' industry was even more concentrated than it is now, but barely significant in the whole economy. In the 1930s less than 1 per cent of the population was employed in industry. Of this, 80 per cent was located in the coastal cities.

One component was a legacy from the late-nineteenth century in the form of enterprises with some strategic interest, such as ironworks (Wuhan), shipyards (Shanghai), arsenals (Jiangnan, Anqing), and their associated coal mines (Kailuan, Pingxiang) and railways. These had mostly been developed with the help and participation of regional viceroys and government officials in order to strengthen China's military capability.

A second component consisted of rapidly growing light industries (notably cotton-spinning, textiles, and flour-milling) in the foreign-controlled Treaty Ports and surrounding areas. Although considerable foreign (Japanese and British) capital was invested in these industries, they were mostly controlled by a growing class of Chinese entrepreneurs. One reason for their location was that the Treaty Ports, with their comparatively free and stable political environment, were financial, marketing and export centres and therefore sources of capital and outlets for the products of industry. The textile industry grew at a sufficient rate that by the 1930s imports of cotton goods, formerly the largest single item, had been almost eliminated.

A third part of the Communist government's post-war industrial inheritance was the Japanese legacy of heavy industry in Manchuria. Japan had deliberately set out to convert the three northeastern provinces of Liaoning, Jilin and Heilongjiang into a source of raw materials for Japanese industry. Having acquired the railway between Changchun (Jilin province) and Dalian (Liaoning) from Russia in 1905, the Japanese mounted a large financing operation to turn it into a form of industrial conglomerate – the South Manchurian Railway Company – with the object of exploiting the northeast's resouces. From 1915 onwards, mines, steel-works and chemical factories were developed on a large scale. Two of China's largest undertakings, the Anshan steelworks and Fushun opencast coalmine, were developed in this period. Between 1932 and 1945, this huge enterprise was subsumed in the Japanese puppet-state of Manzhouguo (Manchuria). On reversion to Chinese control in 1945, the northeast formed China's only significant heavy-industrial centre. With the Soviet evacuation of the area following its defeat of the Japanese forces there, the industrial complex fell into the hands of the Chinese communists and, administered on Stalinist, centralised lines, was an important influence on their later strategy.

Finally, a smaller category of industry inherited in 1949 lay in the interior of the country. This was virtually all of recent origin, having been relocated as the Japanese occupied eastern China in 1937–39 and the government sought refuge in Chongqing (Sichuan province). Engineering and steel enterprises were set up in Sichuan, Gansu and Shaanxi.

Changes after 1949

Following the Communist Party's accession to power in 1949, far-reaching changes took place in the control, management and objectives of industry. These changes had considerable geographical impact.

First, in 1954–5, capitalist-run enterprises were placed under 'joint state–private' ownership. In effect, their owners were compensated in bonds for their takeover, mainly by the local authorities. In 1955–6, individual-controlled enterprises – mainly in commerce – which had represented nearly two-thirds of national income, were converted into co-operatives, controlled less directly through state funding, supply and marketing agencies.

At the same time, central planning was initiated. The First Five-Year Plan (1953–7) was centred on 156 large projects, mainly construction of heavy industrial enterprises with Soviet assistance. A wide range of industrial goods and foodstuffs became subject to rationing and allocation by the bureaucracy at fixed prices which gradually failed to reflect relative scarcity.

With the inception of central planning, industrial enterprises came to have a specific function in the government's development strategy. To raise funds for investment, the general policy was to keep raw material prices low and prices of manufactured goods high. The government received funds in the form of profits and taxes on industry. In return, the enterprises – overwhelmingly owned by central and local governments – were allocated investment grants and working capital free of charge.

Spatial resource flows

The system had two main spatial implications. The first was that geographical flows of capital resources became important. Existing industry, mainly in the coastal regions, subsidised the establishment of new enterprises inland through the tax/profit system. A large source of these financial surpluses was light industry (manufacturing consumer products), in particular in the Shanghai area and in Jiangsu (see Case Studies 7.1 on p. 196 and 8.1 on p. 220). During the 1953–7 plan period, Shanghai alone financed 20 per cent of government expenditure. This situation, modified by the development of profitable natural resources (notably in Qinghai and Xinjiang), is still evident.

Concentrations of industry manufacturing highly profitable and highly taxed products have therefore tended to be reflected in provincial and local government surpluses and deficits. But the sources and recipients of investment funds are even more concentrated than figures for provinces or industrial sectors suggest. Large-scale enterprises are notably more profitable than small-scale ones, and within heavy industry some sectors provide a far greater surplus than others. An industrial census in 1986 revealed that a mere fifty enterprises provided 17 per cent of state income; of these, seventeen were petrochemical plants and fourteen were in the power industry.

In 1988 the No. 1 Auto Factory (at Changchun, Jilin province, one of the First Plan projects) and eight cigarette factories provided over 200 million yuan (worth about £33 million at the time) each in profit and tax. Large-scale iron and steel enterprises, such as Anshan in Liaoning and Shoudu in Beijing, were also big contributors. However, the largest of all was the Daqing oilfield (Heilongjiang), at over 3 billion yuan a year.

Location and self-sufficiency

The second implication of the central planning system was the removal of decision-making powers from individual enterprises, including decisions about siting, sources of supply, and markets. Central planning was also not all that 'central'. Provincial-level and city governments were prone to duplicate industries. Being involved in running enterprises, each preferred local sources of supply. This meant the setting-up of establishments of less than efficient size.

This 'local centralisation' of decision-making and a tendency to local 'empire-building' led to irrational distribution of industries and high costs. A 'prestige' industry such as iron and steel might be represented in some cities by several enterprises run by national, provincial, and city governments, all competing for supplies and unable to find adequate markets. Heavy industries, supposedly the hallmark of an advanced industrial country, have been over-represented and frequently unprofitable. As an example, twenty-four of the twenty-nine provinces produce motor vehicles, but only six produce more than 20,000 a year. The reality of central planning, with its reliance on a devolved yet bureaucratic management system, was therefore very different from the intended efficiency and avoidance of duplication.

Reinforcing the bureaucratic impetus towards small, self-sufficient economies across the country has been an influential current of thought within the party – and especially the army – seeking the same ends for strategic reasons. Partly conditioned by the circumstances of the 1930s, when industry in the hinterland helped to hold off the Japanese invasion, and by the Communist Party's own experience in building up strongholds against Chiang Kai-shek, this was a response to apparent threats first from the United States, then from the Soviet Union after the Sino–Soviet split of the early 1960s. The bulk of factories set up with these reasons in mind was built between 1965 and the mid-1970s in Sichuan, Guizhou, Yunnan, Shaanxi and western Henan, Hubei and Hunan, and numbered around 2,000 large or medium-sized plants (see also Chapter two). Most of these sanxian ('Third Front' as opposed to the coastal 'First Front') plants are involved in military-oriented aerospace, electronics, chemical manufacture and engineering. Many are in isolated locations and have inadequate energy supplies and transport services; in 1986 a plan was announced to alter their production to include civilian goods and relocate some of them nearer to convenient communications.

Local self-sufficiency has also been reinforced by a combination of devolved locational decisions and irrational centralised priorities, resulting in a lack of investment in transport. Seen as a 'non-productive' sector, and therefore not high on the list of priorities for investment, road and rail facilities have consistently lagged behind industrial development. A response of local governments and industrial departments has therefore been to minimise transport needs by locating plants near to raw material sources or near to urban markets, even at the expense of relocating labour (often building accommodation for the workforce on site) or polluting city environments and using scarce agricultural land.

The spatial concentration of industry and regional specialisation in production have been reduced in the past few decades, though at a high cost. Figure 7.1 gives an indication of the differences in output of industry across the provinces in 1985. It shows that the earlier industrial base areas, which proved a vital source of funds and expertise for new enterprises

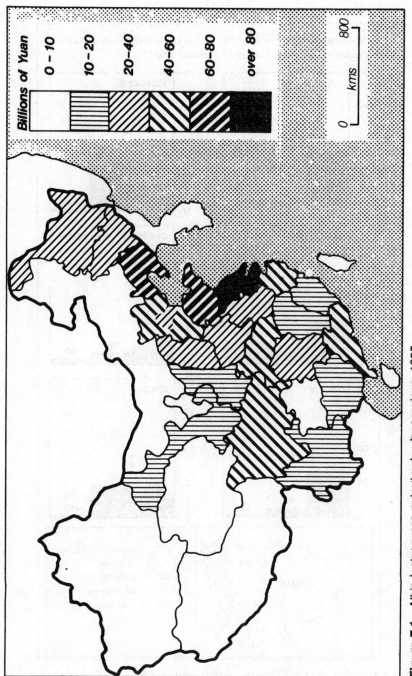

Figure 7.1 All-industry gross output value by province, 1985
Source: China Stasical Yearbook 1986

Billions of Yuan

0 – 10

10 – 20

20 – 40

40 – 60

60 – 80

over 80

0   kms   800

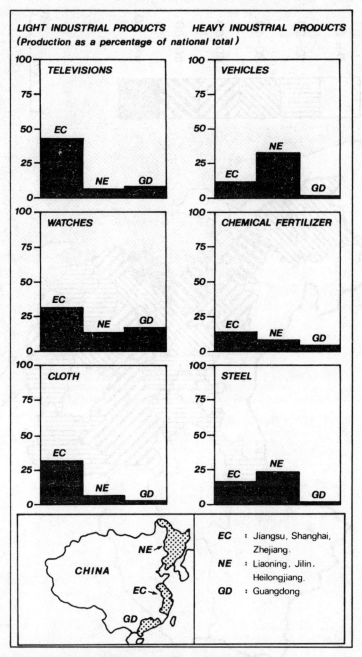

Figure 7.2 Industrial specialisation within coastal China
Source: China Statistical Yearbook, 1986

across the country, have shown themselves to be remarkably durable (as an example of the attraction of new investments, see Case Study 7.1 on Shanghai).

Other significant spatial variations include the differences between the northeast and the east coast in the production of heavy and light industrial goods (indicated in Figure 7.2 for six products), and the gap between Shanghai and Beijing (for instance) and inland provinces in the average size of enterprises (in value terms), as shown in Figure 7.3.

Effects of reform

Since 1979 the economic reforms have had increasing impact on the nature and location of Chinese industry, away from the previous influence of central planning, hierarchical bureaucracy and strategic paranoia. Industrial reforms (or as they are often termed in China, the urban reforms) have essentially operated on two aspects. The first is in altering the nature of enterprise management. Enterprises have gradually become more autonomous units, taking their own decisions about production and investment. Managers have taken over most of the powers of party officials, who previously could dominate the running of factories. No longer is all of a firm's output subject to planning quotas. The relationship between the enterprise and government authorities increasingly takes the form of a contract, under which profit targets rather than physical output criteria are stressed. Smaller state enterprises are now often leased out to co-operatives or individual managers.

The second area affected by reform is the economic environment within which enterprises must make decisions. Prices are gradually being made to reflect comparative scarcities. This applies both to actual products and to investment capital. As part of the reform programme, free grants of money to enterprises by the state have been gradually replaced by bank loans. Ever greater numbers of enterprises are impelled to seek profitable projects and face up to the increasing proportion of production able to be sold at free-market prices. They are now facing charges for their use of capital (interest on loans or bonds) and able to retain more of their profits to finance their own investment (with taxes replacing the surrender of profits to the state).

Because the state now permits enterprises and administrative units to invest in each other, this has become an important outlet for firms seeking profits with any surplus funds. This has broken geographical ties on industry and enables enterprises to seek out the most cost-effective sources of raw materials and most profitable markets for products. By early 1987, 32,000 'horizontal economic ties' (supply or processing contracts, investment agreements, etc.) had been registered with state commercial and industrial departments.

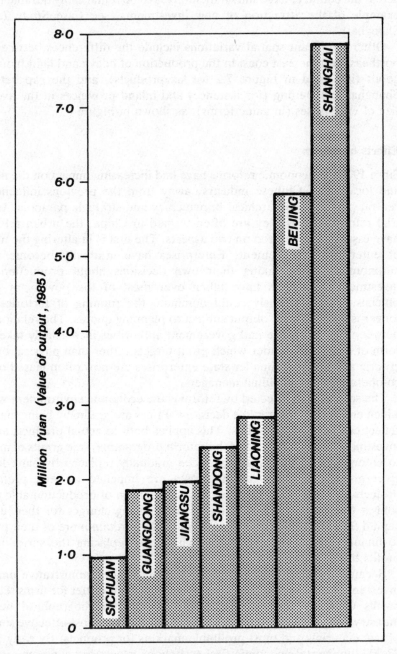

Figure 7.3 Average size of industrial enterprises (by output value) for selected provinces

Source: China Statistical Yearbook 1986, Chinese edition. p. 282.

Another approach to bypassing administrative boundaries, during the period when complete enterprise autonomy has not yet been achieved, has been to set up 'inter-regional horizontal economic networks', essentially regional association agreements between local authorities setting up clearing-houses for firms seeking funds, supplies or markets (see also Chapter two). By 1987, twenty-four of these were in operation. They fall into three categories:

1 *Province-level associations*, such as the Northeast China Economic Zone (Liaoning, Jilin, Heilongjiang provinces and eastern Inner Mongolia, set up in 1983), and others covering north, southwest, south, northwest China, and Shanghai (including Jiangsu, Zhejiang, Anhui and Jiangxi provinces);
2 *Interprovincial associations of lower-level authorities* such as the Bohai Economic Zone (centred on Tianjin) and the Xuzhou Zone;
3 *'City economic combines'* (a form of twinning of non-adjacent urban centres – for instance Shanghai, Nanjing, Wuhan and Chongqing on the Chang Jiang [Yangzi] River), and cities along the Longhai Railway).

Agreements covering compensation trade (barter) between local authorities have emerged from these associations. They are also institutions within which planning co-ordination on economic matters – transport links, development of energy resources, environmental protection, etc. – is increasingly carried out. They frequently lead to agreements between enterprises to form large industrial combines, whose vertical integration (from raw material supply to assembly of final products) enables them to guarantee continuity of production in the more uncertain environment which the reforms have caused.

As a result of these linkages, economic decisions are gradually being separated from administrative ones. Decisions are increasingly devolved to enterprises themselves, or to supervisory bodies whose boundaries make more sense than those of the local authorities.

Regional development

Until the beginning of industrial reforms in 1979 there was no explicit regional policy aimed at increasing the incomes or level of development of poorer areas. A series of measures within the central planning system, involving concealed subsidies to industry and agriculture, were the only attempts to perform this function. These operated by means of nationwide fixed prices for industrial products and specially reduced prices for agricultrual inputs (fertiliser, insecticide, etc.) for poor areas.

Since 1979, the government has set its face against subsidies, arguing that the poor areas (mainly inaccessible mountain areas in the east and centre, plus some parts in the west of the country) will benefit from greater growth in the coastal region. Better existing and planned infrastructure (roads, railways, telecommunictions, education) in many coastal areas mean that foreign technology and management will be more easily absorbed through joint ventures, investment and trading arrangements there. Industrial productivity is expected to increase differentially across the country, declining from east to west and from coast to interior. No longer is the coast seen as particularly vulnerable to military threat, given the government's reassessment of the prospect of war and the new weaponry likely to be involved. This strategic factor has been further reduced in its impact because of a decline in military influence on politics.

While coastal development may be an efficient way to use scarce investment funds, it raises problems of its own. For incomes and standard of living to increase acceptably in the poorer areas in the longer term, solutions must be found to enable new techniques to filter through from the more advanced east of China. Some form of long-distance co-operation agreements, of a similar type to the regional associations mentioned earlier, may help. Local authorities in the east (six provinces plus the cities of Beijing, Shanghai and Tianjin) are pledged to invest in Tibet, Qinghai, Xinjiang, and Guizhou, and to supply technicians to set up factories, build communications facilities and power stations.

In exchange, the eastern provinces receive raw materials (nonferrous metals such as aluminium and nickel, and coal) and processed foodstuffs. These compensation trade agreements can be seen not as an altruistic gesture or a pure regional development strategy, but as a way of realising regional comparative advantage while bypassing so far poorly developed commodity and investment markets. As pressure is put on central government finances, they are also a way of ensuring that funds in the hands of the provinces or individual enterprises are used for investment in infrastructure rather than being diverted to excessive consumption.

Nevertheless, this semi-commercial approach to regional development is inadequate to satisfy the demands of people in the interior whose localities may not be significant sources of raw materials (and therefore less likely to benefit from a trade approach), and who see themselves left behind by growth in the coastal cities and productive agricultural areas. With greater freedom for political expression within the party and National People's Congress, the government has had to consent to explicit regional aids. A budget allocation of 500 million yuan has been made since 1980 to areas below the defined poverty line (receiving less than 200 yuan of income and 200 kg of grain per year per person) covering just under 9 per cent of the rural population. In these areas, new rural industries are allowed income (turnover) tax exemptions (typically for five years) and product tax 'holidays' (a delay in liability for tax of three years). Since 1986, cheap infrastructure loans have also been made available to such areas.

These measures of agricultural support and incentives to expand rural industry may indeed be the minimum necessary if industrialisation in the rural areas is to take place in order to absorb local labour. Before 1978–9, rural industry was promoted (mainly owned as co-operatives and based in communes and brigades). These were intended to improve productivity in agriculture by making farmers' tools and other supplies. Now industry is seen as a way of absorbing surplus rural labour and preventing migration to the cities and as a way of raising incomes in poor farming areas where the agricultural reforms are having little effect. It will therefore become more oriented towards producing consumer goods for the towns, and supplying components to urban industry. Its growth will therefore be a feature mainly of urban peripheries.

Small-scale 'rural' industry, based on subcontracted work from large city factories, has so far been developed furthest in the hinterland of Shanghai (e.g. in southern Jiangsu – see Chapters six and eight), and also in Guangdong. In this latter province, the process is benefiting from land and labour costs lower than those in Hong Kong and is, in effect, an extension of the Hong Kong economy just across the border. Such rural industry is the fastest-growing category of industry (at two and a half times the average industrial growth rate), and it will undoubtedly play an important role in the future of the Chinese economy.

ENERGY

China stands apart from other developing countries not only in the structure of its industry and demography, but also in its pattern of energy consumption. At the heart of this is the country's overwhelming reliance on coal as a source of fuel. Of total energy use, including household fuels, coal makes up around a half. The remainder is either biomass (wood, dung and straw, or 'bio-gas' [methane] derived from them), forming 25 to 27 per cent of the total, oil or natural gas. In terms of primary commercial (traded) energy, the preponderance of coal (75 per cent) is even greater. Oil (20 per cent) and natural gas (less than 3 per cent) are well behind. This feature contrasts sharply with other developing countries, where wood or oil are more commonly used.

Use of these sources of primary energy also differs from most other countries. Of energy used by final consumers (including industry), electricity only accounted for 18 per cent in 1980, a much lower proportion. Industry consumes a large part (over half) of coal, oil, and gas, and uses around three-quarters of electricity generated. Of electricity used by industry, a disproportionately large amount is consumed by heavy industry, especially in steel and chemical manufacture. Overall, compared with other countries, China is a very high user of energy for the size of its economy. Partly this is due to the high share of industry in the economy – a result of government policy – and partly to low efficiency in industrial use of energy. It is also explained by the low price of coal (one-fifth to one-sixth of the world average), oil and electricity, which has removed a good deal of the incentive to use energy efficiently.

In addition to the problem of efficiency, China's reserves of coal, though large by world standards (1,440 billion tons, 13 per cent of the world total), work out at only 40 per cent of the world average in terms of immediately exploitable reserves per capita. It also has to be remembered that hydrocarbons form attractive raw material sources for important products in chemicals, especially for fertilizers and artificial fibres, and to some extent there is competition for oil and coal for non-energy uses.

Problems of demand

With continuing economic growth at present rates, coal output would need to approach 2 billion tons by the year 2000 if existing patterns of use are also maintained. Yet it seems that the maximum that it will be possible to mine and move at that time may be around 1.4 billion tons, unless mine development and transport facilities are expanded faster than at present. Scope for efficiency improvements in the use of coal is limited. Significant gains since 1980 have already been made by reducing the role of heavy industry and improving technology. Scope for using alternative fuels is also restricted. Per capita oil and gas reserves are assessed at around half the world average.

Demand for alternative fuels is also rising. Rural biomass use is running at the equivalent of about 220 million tons of coal per annum, causing degradation of soil and loss of animal fodder, yet still being 20 per cent short of demand. Proliferation of motor vehicles is greatly increasing demand for oil, yet oil production is only growing very slowly, as the main producing area (Daqing in Heilongjiang province) gradually runs down and major new fields have to be found to replace it. Because of all these factors, reliance on coal is likely to continue in the foreseeable future.

Problems of location

Distribution of China's fossil-fuel resources has only made these pressures restraining economic growth more difficult to bear. Indeed, energy resources as a whole are concentrated in a perverse way which serves to exacerbate the burden of the great divide in the country between coast and interior. Nearly three-quarters of proven coal reserves lie in north China (in Shanxi, or on the Ordos plateau in southern Inner Mongolia and northern Shaanxi). Other sources of coal lie in Liaoning and at Kailuan near Tangshan (Hebei province). Major deposits of lignite are located on the grasslands of northeastern Inner Mongolia.

By contrast, around 70 per cent of the available potential for generating hydro-electric power in large-scale projects is found in the southwest, far from the main areas of electricity consumption. The steep upper reaches of the Chang Jiang (Yangzi) tributaries (the Dadu, Yalong, and Jinsha in Sichuan and the Wujiang in Guizhou) and the Lancang (Upper Mekong) in Yunnan are prominent possible locations. Another, equally remote, concentration of hydropower is the Upper Huanghe (Yellow River) in Qinghai, Gansu, and Ningxia.

The main areas of power consumption, however, have been in the east of the country. In thirteen eastern provinces, energy production accounts for only 13 per cent of the national total, while they consume 65 per cent.

Because of inadequate investment in mining, transport, and in particular in electricity generation facilities, these disparities in resources have resulted in dramatic impediments to development. In China as a whole, official estimates suggest that electricity supplies fall short of demand by nearly 10 per cent. In the southeastern provinces alone, there is reckoned to be a shortage of no less than a third. As a consequence, there are frequent power black-outs, quotas for new connections, and industrial equipment unable to run at full capacity. It is also reported that the incidence of black-outs has been spreading westwards, indicating that the situation has been worsening. Even in the northeast, where heavy industry consuming coal and electric power is cheek-by-jowl with big coal-producing centres, there have been continuing shortages.

The problem has largely resulted from decisions taken over a period of thirty years and with a different view of priorities. For a long time, the response to a rising demand for energy was to try to restrict it by keeping tight control over central government spending on supply facilities. The state encouraged reliance on local sources: small hydro-electric stations, serving a village or two, are still a typical feature of China, and supply over 40 per cent of electricity used in the countryside (see Case Study 7.2 on p. 200). Methane-producing bio-gas pits were also promoted, but their success is mainly restricted to rural Sichuan. Development of regional grids was very slow. Networks covering the whole of north China and the northwest were only set up as late as 1980. Until the reforms of 1978-80, self-reliance in energy was encouraged every bit as much as self-reliance in investment funds or in production of goods.

Oil and natural gas

While pressure on energy supplies has been a constant feature of China's development, the greatest change in circumstances since 1949 has happened in relation to oil. Traditionally an important consumer of foreign oil, China became overnight (in 1963) a potential exporter thanks to the discovery of one oilfield. The Daqing field is located in the central Songhua–Nenjiang plain in Heilongjiang province, and although its production has now peaked, it still contributes nearly 40 per cent of the country's petroleum.

With the ready availability of domestically produced oil, and – at least initially – a low demand for its use as fuel, a number of important options were open to the planners. One was to tackle the pressure of rapidly rising population on the limited agricultural land in the country. Here, oil could be used as a chemical feedstock to produce plastics for use in protecting crops, or to make chemical fertilisers. Alternatively, it could be used to manufacture artificial fibres, replacing cotton needed for clothing and other textile products, and allowing the land released to be used to grow food.

Another option was to sell oil on the world market, and with the foreign exchange import technical equipment to raise industrial productivity. In practice, both strategies have been employed to varying degrees at different times. Oil exports began in 1973, and helped to pay for a large programme of petrochemical plant imports, mainly aimed at producing fibres and plastics (see Case Study 9.1).

This development greatly reinforced the importance of the oil-producing areas. Since the discovery of Daqing, smaller fields have been located around the shores of the Bohai sea, (Liaohe, at the mouth of the Liao River, Dagang near Tianjin, and Shengli, at the mouth of the Huanghe River in Shandong) as well as offshore (near the Dagang field). Petrochemical plants were set up at Daqing, Niaoyang (Liaohe), and Shengli, as well as at locations closer to consuming centres (Yanshan near Beijing, Jinshan near Shanghai, and Nanjing). In addition, refining capacity already existed at Dushanzi close to the Karamai oilfield in Xinjiang, opened in the 1950s, and at Lanzhou. By the end of the 1970s, nearly all these oilfields and processing centres had been connected by pipelines and linked to oil terminals at the ports of Dalian (Liaoning), Qinhuangdao (Hebei), and Qingdao (Shandong).

By 1976, however, the main field of Daqing had reached its peak output. From then on, its production (at about 50m tons per year constituting half of the national total) was restricted to that level. As a result, pressure from rising demand became acute. During 1977–9, the authorities were seized with wild optimism about the extent to which oil could be discovered and tapped, and concluded a long-term trade agreement with Japan to import technological equipment in exchange for oil and coal shipments. When this was found to be unrealistic, the government turned to the prospective offshore oil reserves (forming a third of the estimated 30 to 60 billion tons of reserves) and to the use of foreign capital to develop them (see Chapter nine).

So far, the most successful areas have been the Yinggehai basin (mainly for natural gas) off southwest Hainan Island, the Pearl River estuary, and extensions of the Bohai oilfield. These discoveries have led to major developments in the areas serving them, but as yet contribute very little to total production (see Figure 9.6 on p. 246).

Another recent development has been programmes to replace coal briquettes by gas for domestic heating and cooking in the large and medium-sized cities. The gas originates as a by-product of oil extraction (as in the case of Tianjin) or from coal (Shanghai). The objective is to reduce pollution and demand on transport facilities.

Natural gas already forms a significant part of the local economy of Sichuan, an important gas-bearing basin. There, it is used not only as chemical feedstock (notably at the chemical fibre complex at Changshou) but also as fuel for road vehicles, sometimes carried in large bags on top rather than in pressurised cylinders.

But a steady development of onshore oilfields (Huabei in Hebei province, and Zhongyuan in Henan, for example) has compensated for Daqing's inability to supply rising demand. Development of the Shengli oilfield should shortly enable it to match Daqing's output. Fortunately, the new fields are within easy reach of centres of consumption in central and north China.

Small oilfields maintain local supplies to the more inaccessible areas of the country, as for example the Qaidam fields in Qinghai, and Yumen in Gansu (the oldest commercial oilfield in China). Oil-shale is processed at Fushun in Liaoning and Maoming in Guangdong. Because of state-fixed low prices for crude oil and high prices for oil products, many local authorities have sought a slice of the profits, and refining is therefore a very dispersed activity. Most of these fields have their own refineries.

Since 1980, central government regulations have banned the use of oil as fuel in power stations or industrial boilers, saving an estimated 20 per cent of oil consumption. Together with continued discoveries, this has allowed exports of crude oil to be increased to over 20 per cent of production. This measure has, however, increased the stress on coal supply, especially to east and south China.

Coal

Since coal features so prominently in China's resource endowment, and since the country is a significant producer on an international scale (now the world's largest), decisions taken about this fuel have had an immense importance for the nation's economy.

During the last three decades the tendency has been to seek a maximum rate of extraction for a minimum of investment. This has applied to both coal and oil. Despite this, or more properly because of it, the rate of growth of output has been low, barely keeping ahead of economic growth. Even though investment in mine development has been greatly expanded since 1979, growth in coal output has slowed.

Dramatic slowing of growth in oil extraction (to around 3 per cent a year) has led the authorities to look for ways of managing the consumption of this increasingly precious resource. One response, as we have seen, was to switch to using coal wherever possible, and oil has been phased out of low-grade fuel uses during the 1980s.

Not only has fuel policy been changed, but official attitudes to the use of oil and coal as chemical raw materials have also undergone some rethinking. With the improvement of agricultural productivity a key priority, production of chemical fertiliser has become a crucial element in economic growth. During the 1970s when oil seemed relatively abundant it was developed as a feedstock for producing nitrogenous fertilisers (urea- and ammonia-based compounds). But it has now become increasingly unattractive to use it in this way. The focus of attention has therefore turned to the use of coal as a raw material.

Since the siting of fertiliser plants is not heavily dependent on the supply of more than one raw material, many of these have been developed in coal-producing areas. Shanxi, being by far the largest coal-producing province (with nearly a quarter of national output) is being developed as a centre of fertiliser manufacture. The country's largest nitrophosphate complex was established in 1986 using German technology at Lucheng, near Changzhi in the southeast of the province. Coal gasification plants (to supply urban gas) and liquefaction plants (to make fuel oil) are also under construction. Where coal has a use purely as a source of energy, priority is being given to developing it in areas where other raw materials are located. The Liupanshui mines in Guizhou have been developed in conjunction with the Panzhihua iron mine and steelworks in Sichuan. Aluminium, a very energy-intensive metal to produce, is the subject of a big development plan involving sites not only near to hydro-electric resources, but also in locations where coal and bauxite deposits are close together (such as Jiaozuo in western Henan).

Only a small proportion of coal is washed to remove rock and impurities (around a fifth of output, mainly from mines run by the central government), and therefore efficiency in transporting and using it is relatively low. This led the government, from the mid-1970s and with increasing urgency, to look to using coal in large-scale modern power stations located at the sites of major mines. A number of these are now in operation in Shanxi (Shentou, Datong No. 2), at Tangshan in Hebei (Douhe), at Pingdingshan in Henan (Yaomeng), at Yuanbaoshan in Inner Mongolia, and elsewhere (Figure 7.4).

More generally, the trend towards localised use of coal is leading to a reconcentration of heavy industry in mining areas, away from the coastal centres of population devoid of energy resources. Price changes under an increasingly market-oriented production system will undoubtedly accelerate this tendency, as coal production ceases to be subsidised and transport

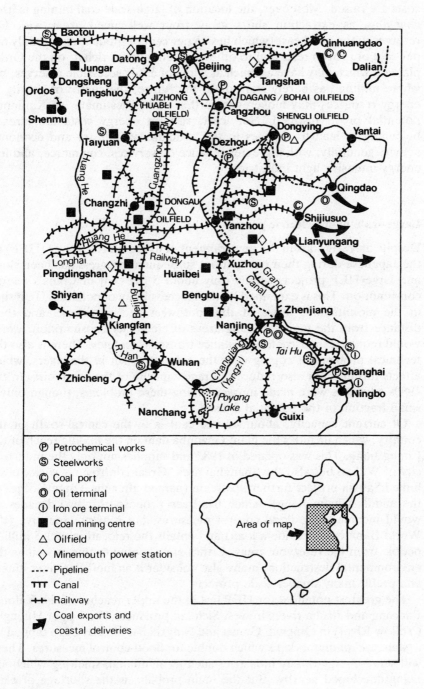

Figure 7.4 Energy and transport in east China

costs are raised. Moreover, the location of large-scale coal mining is itself changing, as extraction shifts away from well-mined areas with low reserves in the northeast (which has 15 per cent of production but only 6–7 per cent of proven reserves) to others, potentially far richer, on the Ordos plateau, which may have as much as a half of easily accessible reserves, but where mining has barely started. In the future, the centre of gravity of energy resources may move even further west, if estimates of Xinjiang's potential prove justified. Not least, rises in energy costs are already beginning to lead to changes in the structure of industry and economic activity generally, with a greater reliance on services, commerce, and less energy-intensive light industries.

Large-scale hydro-electric power

Despite an early start in the development of hydro-electic power (HEP) by the Japanese during their occupation, subsequent progress has been slow, and large HEP projects supply only about 5 per cent of China's energy consumption. This is explained by the inaccessibility of potential HEP sites in the mountainous areas of the southwest and northwest, and their distance from the major existing areas of electricity consumption, which would require expensive long-distance transmission lines. There is also the technical problem of coping with the amount of silt in the rivers, which affects turbines but also reduces storage capacity of the reservoirs. In the 1980s advances were made in overcoming these problems, though only a small fraction of the potential is used.

Of current capacity, about 40 per cent is in the central-south of the country, which includes the giant Gezouba dam on the middle reach of the Chang Jiang. This was opened in 1982 and supplies not only the industrial city of Wuhan but also the Shanghai area. Great controversy surrounds a linked Sanxia project further upstream (named after the Three Gorges of the middle Chang Jiang), which has been proposed for two decades. It would include some flood control measures, but is very expensive (the World Bank may provide a loan) and entails the relocation of 1.5 million people from the reservoir areas. Critics are opposed to this as well as the environmental destruction; many also consider it an inefficient investment and prefer many smaller-scale projects.

The greatest potential for HEP lies in the upper reaches of the Yalong, Lancang and Jinsha rivers in west Sichuan province. The Upper Huanghe (Yellow River) in Qinghai, Gansu and Ningxia has already been tapped by a 'staircase' of major dams which double for flood-control measures. These will serve energy-thirsty industries such as aluminium-smelting which are being developed nearby. But the main problem is the shortage of electricity in already developed areas, especially in the east. Here there are attempts to promote HEP development, as continuing power cuts urge on the search for alternative power sources.

Other energy sources

With energy supply a limiting factor in China's development, and with dramatic regional variations in its availability, alternative sources have been seriously sought since the late 1970s. Particularly noteworthy is the development of geothermal fields, which are quite widespread throughout the country. Some factories and residential districts in Beijing and Tianjin are heated with water from local wells. An experimental geothermal power station at Yangbajain in Tibet supplies most of the electric power used in Lhasa.

Other low-intensity sources, such as domestic methane production from manure (bio-gas), continue to be encouraged in rural areas where electrification is expensive, though it is restricted by low winter temperatures which slow the process. Use of wind and tidal power is still at an experimental stage, although several small installations have been built in Tibet, Qinghai, and along the east coast.

Nuclear power is not yet developed, but a series of plants are under construction or planned in the coastal areas with the worst power deficits. The first, at Qinshan in Zhejiang (300 MW) is due to enter service in 1989. It will serve Shanghai and the east China grid. Another at Daya Bay in Guangdong (1,800 MW) is being built as a joint venture with Hong Kong power interests and will serve Hong Kong and Guangdong. Others will be located in the Shanghai area on the south bank of the Chang Jiang (Yangzi) near Changzhou and at Jinshan, on the coast south of Shanghai. Yet another is planned for the coast of Liaoning. By 2050, nuclear power plants could, on ambitious estimates, provide half of China's electricity.

TRANSPORT

Large disparities in natural conditions, and in the levels and types of economic development superimposed on them, might be expected to put a premium on the possession of an effective transport system. With such a large part of China's natural resources located in the west and centre of the country (e.g. metal ores, cement) or north and northeast (especially coal), and such a proportion of the population living along the coast, there is inevitably a great requirement for transport of bulk goods. Concentrations of certain types of industry – for example, heavy engineering in the northeast and textiles in east China – increase transport demand even more. Added to this, climatic variations, generally from north to south, create demand for long-distance transport of agricultural products.

The structure of the economy, too, has been geared towards heavy use

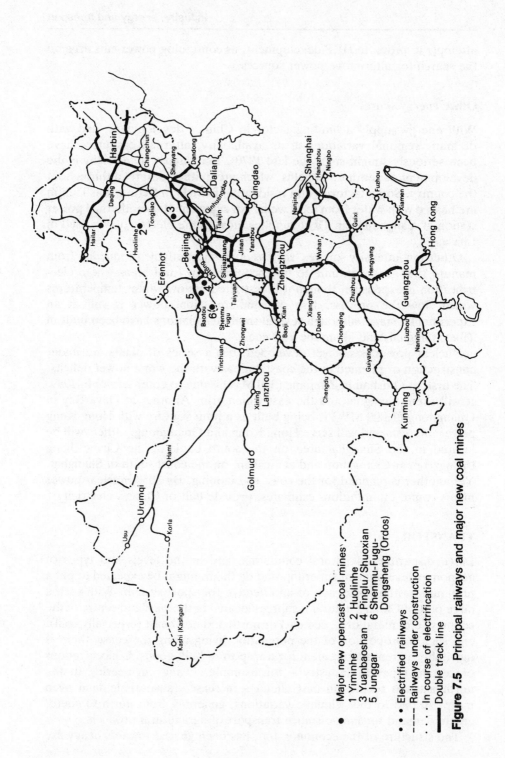

Figure 7.5 Principal railways and major new coal mines

Major new opencast coal mines

1 Yiminhe   2 Huolinhe   3 Yuanbaoshan   4 Pinglu/Shuoxian
5 Junggar   6 Shenmu-Fugu-Dongsheng (Ordos)

Electrified railways
Railways under construction
In course of electrification
Double track line

of transport. One apparent cause has been the special stress on heavy industry combined with economically irrational siting of enterprises. As a result, China's use of transport in comparison with the size of the national product is outstandingly high in world terms.

Railways

A feature of the country's transport activity is the extent to which it is dependent on the use of railways. Rail accounts for nearly two-thirds of freight traffic. This is despite the fact that, for the country's size, its rail network (at 53,000 km) is quite small. Intensity of use of the system is therefore very high – second only to the Soviet Union.

What is more, the railway network is heavily concentrated in those areas where industry has historically been centred (Figure 7.5). No less than 23 per cent of the railways are in the three northeastern provinces. This reflects their intensive development there by the Russians and Japanese before 1945. A further 35 per cent of the system serves the plains of east China and still largely reflects the priorities of the 1890s–1930s when it was built – the transport of armies and officials between a few main centres of population. During the late 1960s and the 1970s a number of strategic railway lines were built, mainly in the southwest, to help relocate military industries and develop the resources of the interior under the 'third line' policy.

Waterways

Second in importance as a means of moving goods is the old-established system of waterways. Just under 20 per cent of freight movement takes place by inland navigation or coastal shipping, and it is therefore quite an important element of the overall transport scene.

Once again, however, there is a heavy concentration in the east of the country. A third of the system of canals is located in Jiangsu, Shanghai, and Zhejiang astride the mouth of the Yangzi. Another 10 per cent is found in Guangdong, centred on the Zhujiang (Pearl River). And, not surprisingly, the main artery is the Chang Jiang (Yangzi) itself, navigable a total of 2,958 km all year round, as far up as Yibin in Sichuan. Together with its navigable tributaries, the Chang Jiang network extends to 18,000 km. Since 1979 the main ports as far as Chonqing have been gradually reopened to foreign ocean-going ships, restoring to the river its pre-war importance as a doorway for international trade. Ships as large as 10,000 tons can reach Nanjing; Wuhan is accessible to those up to 5,000 tons.

Great hopes have been placed on the ability of the waterway network to take the strain from the overloaded railways, and yet it has important features which act as a constraint on extending its use. The first is that a great deal of the system is too shallow to take even deep-draught barges. Only 8,000 km is available to boats of over 8,000 tons. Nearly half the length is less than 1 metre deep. One potential east–west artery, the Huanghe (Yellow River) is so prone to silting that it is only navigable for short, disconnected lengths. The Grand Canal, built initially in the Yuan Dynasty (thirteenth century) to take the Imperial grain tribute from south to north, is only now in the course of reopening. Apart from this and a few waterways in the southwest and Hunan, there is little scope for using this cheap method of transport for north–south traffic.

The second obstacle consists of the large number of dams built across formerly navigable rivers, and designed without shiplocks. After 1949, navigable waterways were expanded dramatically by dredging rivers and restoring canals, yet since 1962 a third of the system has been lost, mainly for this reason. Additionally, a highly seasonal river flow, particularly in the Chang Jiang (Yangzi) and its tributaries, makes important routes unusable at certain times of the year (mainly during late Summer floods). Because of these problems and the historical development of the waterway network, most traffic is over short distances within the coastal provinces.

Other forms of transport

Roads

Roads naturally form the most extensive transport network, with 1 million km, of which two-thirds are located in rural areas. However, they only account for a very small proportion of goods traffic in China (14 per cent of tonne-km in 1985). This is accounted for both by the comparatively small size of the network (China has one-eighth of the road density of the USA) and the low capacity of such roads as there are. Despite a growing programme of road improvement during the 1980s, less than 20 per cent of the length is asphalted, and much of the rest has no kind of surfacing, is narrow, seasonally unusable or clogged with slow-moving agricultural vehicles.

Roads with motorway specifications (dual carriageway and limited access) are only now beginning to be built along some of the main traffic axes in the northeast, east, and south of the country. For most of China's rural areas, however, inaccessibility remains a major obstacle to development.

Pipelines

Since the mid-1970s, pipelines have become important in the movement of oil and natural gas. A network stretches from the Daqing oilfield in the north to oil export ports at Dalian and Qinhuangdao, and onwards to link up refineries and oilfields in Beijing, Shandong and Nanjing. A system in the northwest links the Karamai field with the refinery at Dushanzi and the Yumen field to Lanzhou in Gansu. In total, pipeline usage represents around 4 per cent of total goods movement, and has released useful railway capacity.

Air

Movement of goods by air, though growing rapidly along with general expansion of the network, has yet to make a significant impact. High-value perishable goods – such as Hami melons from Xinjiang – are sometimes sent by air.

Transport of coal

The most notable feature of freight transport is the extent to which it revolves around the movement of coal. China's dependence on coal for energy supply, and the great variations in natural endowment of coal across the country, have already been noted.

Coal movement from the north to the south and east has been a long-standing phenomenon. Over the last few decades, this was compensated for by movement of grain from south to north. However, grain output has grown more slowly than that of coal, and patterns of dependence have changed since the rural reforms were initiated, so that, for instance, more grain is available near the conurbations which consume it, while provinces such as Sichuan have ceased to be important suppliers. Coal movement to the south is therefore a heavily unbalanced traffic – few bulky goods return to the north.

These tendencies have led to a steady growth in the share of coal in transport use. On the railways it now represents over 40 per cent of tonnage loaded, and well over a third of tonne-kilometres carried. It also represents around a third of waterway freight.

The increasing concentration of coal production in the north in recent years (see Figure 7.5), and its emergence as a critical problem for China's overall economic development, have led to major changes in official policy. In 1979, movement of Shanxi coal to other parts of China became a top-level priority, and resources were specifically devoted to projects to aid this. Previously, investment in railway facilities had suffered much the same low priority as other 'non-productive' items. Three years later, planning was put under a Shanxi Energy Base Planning Office, directly answerable to the State Council. This covers north Shaanxi, western Hunan, and central Inner Mongolia (mainly the Ordos plateau) as well as Shanxi province, and represents a comprehensive approach to planning which bypasses departmental and provincial bureaucracies.

Attention is now shifting to the improvement of rail links in eastern China, as the urban economic reforms lead to increased demand in the developed urban areas of the country. However, shortage of land in the productive agricultural eastern areas, as well as shortage of capital funds, has spurred a search for cheaper ways of transporting energy. One solution has been to construct high-voltage transmissions lines from power stations sited on coalfields to important areas of consumption. The high voltage enables transmission losses in the cables to be kept down. To reduce loss even further over long distances, and for other technical reasons, DC rather than AC can be used. So far, high-voltage (500 kV) AC lines have been installed within the regional grids linking power plants and large-scale industrial complexes such as Pingdingshan and the Wuhan steelworks (Hubei), and Yuanbaoshan and the Liaoyang chemical centre in the northeast. An inter-grid system of nearly 5000 km is also to be completed by 1990. A DC line 1,080 km long joins the Gezhouba hydro-electric station on the Chang Jiang (Yangzi) with Wuhan and Shanghai.

To some extent, electric power transmission is beginning to alleviate the need to transport coal over long distances. However, this need will undoubtedly increase in coming years. Interest in substituting electricity for direct burning of coal is also reflected in changes in energy use in transport itself. Until now, steam locomotives have performed most work on the railways, but electrification is now favoured.

Transport of materials other than energy is, of course, also important. The extent to which the various modes of transport are used depends a great deal on the distance between origin and destination. Grain and agricultural produce do not feature very much in rail-transport statistics because, compared with coal or other minerals, provinces tend to be relatively self-sufficient, and haulage distances are short. Grain and vegetables tend to go by water (average distances travelled by a ton around 200 km) or road, rather than rail (average distance over 600 km). To encourage use of roads, transport by rail of goods over distances less than 200 km is now discouraged by high freight prices.

To spare inland railways and roads as much as possible, combined transport and coastal shipping have been encouraged. Grain and coal from the northeast are shipped from Dalian to Shanghai, for instance, and coal from the north is taken to Qinhuangdao and Shijiusuo to be shipped south along the coast. This trans-shipped traffic has in the last few years, together with pressure from coal and oil exports, plus grain and fertiliser imports, created critical bottlenecks at the ports. Specialised loading facilities, container bases and improved inland links, financed in the main by Japan, have been developed since 1979.

A further reduction in pressure on transport facilities can be expected from reforms in the pricing and decision-making powers of industry. Low pricing of some goods (e.g. cement) from centrally controlled factories allows them to compete with plants nearer to the users, with prices on delivery failing to reflect the distance carried or the quality of the product.

Over-use of transport, and use of an uneconomical mode like rail can result.

Passenger services

With reform of China's economic structure, movement of people has become an important issue. Between 1978 and 1985 the use of waterways for passenger transport increased by 74 per cent, railways by over 100 per cent, and bus traffic trebled. Although a great peak in travel occurs at Chinese New Year, when relations visit each other, most of the increase must be due to work travel of one kind or another. Commuting forms an important part of urban passenger traffic but, unlike Western cities, Chinese conurbations have a relatively unfocused commuting pattern. Factories and housing are often combined or close together. Housing, industry and commerce tend to be intermixed, with no clear suburbs. Rail networks in the big cities are not designed to carry short-distance passengers into the city centre.

Another element of work travel is long-distance business travel. Increasing numbers of salespeople or private business people (including peasants selling agricultural produce) are making regular journeys, often quite long, encouraged by regional shortages and the increasing scope given to the free market. Temporary labourers from the countryside working in the coastal cities and Special Economic Zones are also more numerous, adding to the demand for personal travel.

An upsurge of economic activity in urban areas, combined with restrictions on rural inhabitants taking up permanent residence in the cities has undoubtedly encouraged the tendency to travel, either to visit friends or relations or to undertake business. The current replacement of written travel permits issued by employing organisations (work units) with state identity cards is both a recognition of this trend, and a way of making travel easier while keeping control of residence. Once again, pricing policy is discouraging short-distance use of public transport, while increasing amounts of money are spent on high-capacity mass-transit railways and urban freeways.

With the prospective introduction of market-based land pricing in urban areas, it is likely that concentrated central business districts will develop, and along with them more intense commuting patterns. The extent to which the cities – particularly the coastal cities – can fulfil the functions of business and service centres is increasingly seen as being dependent on their endowment in transport infrastructure. As resources are devoted to this and to the other elements of urbanisation, the repercussions on other areas of the economy and other parts of the country will be profound.

CONCLUSION

This chapter has attempted to give an overview of the factors which have influenced China's development policy for industry, energy and transport, and how these have worked themselves out geographically. We have seen how important were strategic reasons, rather than egalitarian ones, in development of the interior by the state. We have noted aspects of transport and resource location which have set limits or imposed heavy costs on such development, and how a strategy of self-sufficiency can be seen as a response to these costs.

Part of the revolution in attitudes since 1978 has been the virtual abandonment of attempts to fly in the face of these penalties, and to opt instead for an unequal geographical spread of development, and an explicit strategy of growth poles based on major cities in the east of the country, notably the fourteen coastal 'open cities' designated in 1984 (see Chapters two and nine). State spending on industry and transport has switched to east China and the coal-producing areas in the north.

But we have also seen how decision-making on industrial location, supply and markets is decreasingly the concern of central state organs and increasingly that of individual enterprises. This is the second aspect of the reforms to have a crucial bearing on China's economic geography. New hierarchies of activity, patterns of urbanisation and spatial distributions of industry and services are bound to emerge as enterprises respond to prices which reflect scarcities of resources, notably land, energy and transport capacity. A possible result may be more intense concentration of financial (and therefore increasingly political) power in a small number of coastal centres, almost certainly Shanghai and Hong Kong. Industrial development may stretch out from these along the coastline, most of which is as yet undeveloped. Pressure to migrate to the growing cities may prove even harder to control, as bureaucratic restraints erode. How far the state remains capable of balancing these tendencies by transfers of resources to the interior, where ethnic minority peoples in sensitive border areas may feel resentment, may be a key question for the future development of China.

How far it is capable of ensuring that industry and commerce pay the price for encroachment on east China's vital agricultural land may be no less important. Increasing pressure for widespread use of private cars may put excessive stress on land-use and energy priorities. The authorities have gambled that greater inequality will allow greater prosperity. They have gambled that greater prosperity will eventually enable population growth, the background against which all China's development problems must be seen, to be halted.

If unconventional solutions, whether a coal-based economy, a draconian birth-control policy, or a radical redefinition of official ideology are adopted, it should come as no surprise. But whatever change occurs in the future, it will almost certainly be felt most acutely along China's maritime face to the outside world.

FURTHER READING

Anon (1984) *China Handbook Series: Economy*, Beijing: Foreign Languages Press.
  A comprehensive and concise factual compendium. Somewhat out of date (the latest statistics are for 1980) and, like most Chinese sources, short on analysis, but a good source book with occasional critical inspiration. This is recommended more through a shortage of books on China's spatial economy since 1978 than for its outstanding virtues as a source.
Dorian, J. P. and Fridley, D. J. (eds) (1988) *China's Energy and Mineral Industries*, Boulder and London: Westview Press.
  This is a collection of essays, mostly by Chinese, which deals less with the availability of energy and mineral resources and concentrates on the planning and policy issues, and on China's connections with trade and foreign investment in these sectors.
Goodman, D. (1989) *China's Regional Development*, London: Routledge.
  This volume contains a number of chapters relevant to the material of this chapter, including transport and energy, and others on spatial aspects of development.
Lieberthal, K. and Oksenberg, M. (1988) *Policy Making in China: Leaders, Structures and Processes*, Lawrenceville: Princeton University Press.
  Examines the evolution of post-Mao policy towards three major energy projects: Shanxi coal, South China Sea petroleum and the Three Gorges dam on Yangzi.
Perry, E. and Wong, C. (1985) *The Political Economy of Reform in Post-Mao China*, Cambridge, Massachusetts: Harvard University Press.
  A useful treatment of the framework within which industry operates can be found in Susan Shirk's contribution (Ch. 8 'The politics of industrial reforms') and Christine Wong's chapter (Ch. 10 'Material allocation and decentralisation: impact of the local sector on industrial reform'). These are not explicitly geographical, but do bring out locational issues, conflicts of interest between coastal and interior provinces, and influences on industrial structure.
Smil, V. (1988) *Energy in China's Modernization: Advances and Limitations*, London: M. E. Sharpe.
  Based on a close analysis of Chinese sources, this considers China's potential energy resources, its role in modernising the economy and the economic and environmental implications of energy policy.
World Bank (1985) *China: Long-Term Development Issues and Options*, Washington: World Bank.
  Undoubtedly the best, most analytical work on China's economy since the reforms began. A general volume is accompanied by several annexes covering important sectors. On subjects covered by this chapter, Annexe 3 ('China: the energy sector') and Annexe 6 ('China: the transport sector') are the most relevant. For a broader view in the context, say, of international development studies, Annexe 5 ('China: economic structure in international perspective') is also useful. The World Bank's central concerns are sectoral rather than spatial, but spatial issues are given separate and considered attention.

Case Study 7.1

SHANGHAI: A MAGNET FOR HEAVY INDUSTRY

Maurice Howard

We have already seen how Shanghai and its hinterland has been important in China's light industry, and in the generation of finance for the rest of the country. The city produces up to a third of many consumer products, being particularly prominent in electronics and metal consumer goods. It has an importance far greater than the size of its population in food processing. Overall, it is responsible for nearly 9 per cent of light industrial production. The size of the city's enterprises is noticeably bigger than the national average (see Figure 7.3), and their productivity in terms of manpower and assets is also greater.

This, combined with the low priority given until recently to investment in the city, has been behind its important role in China's economy. Despite the restrictions placed on its population and finances, Shanghai has hosted a number of heavy industrial developments of national importance. Since the early 1970s these have been boosted dramatically. The city is now an important centre for the production of power equipment, computers, aircraft and motor vehicles.

Two particular enterprises play an important part in the city's economy (Figure 7.6). The Shanghai General Petrochemical Works at Jinshan is China's largest such plant. Built between 1972 and 1977, the first phase of construction was designed to produce 102,000 tons of synthetic fibres (including 80,000 tons of polyester, vinylon, and polyacrylic fibres, a quarter of national output), plus 60,000 tons of plastic resin and 2.8 million tons of other petrochemical products a year. A second phase, built in 1980–3, added a 200,000-ton-per-year polyester chip plant. A third phase on completion will allow the plant to produce a yearly total of 450,000 tons of ethylene. China's current output is about 700,000 tons.

The Baoshan Iron and Steel Complex is Shanghai's second giant industrial installation. The first of three planned phases was built in 1978–85, with a planned capacity of 3 million tons of iron, 3.12 million tons of steel, and 0.5 million tons of seamless steel tubes. The second phase is intended to raise this to 6.5 tons of iron, 6.7 million tons of steel, and 4.22 million tons of rolled steel products by 1990 – more than 10 per cent of national output. A third phase would raise steel output to 10 million tons.

Despite explicit policies of decentralisation, dispersal of industrial activity, population restrictions and removal of urban–rural differences, how is it that Shanghai should be chosen to accommodate two of China's biggest industrial complexes? Other than size, there are several similar

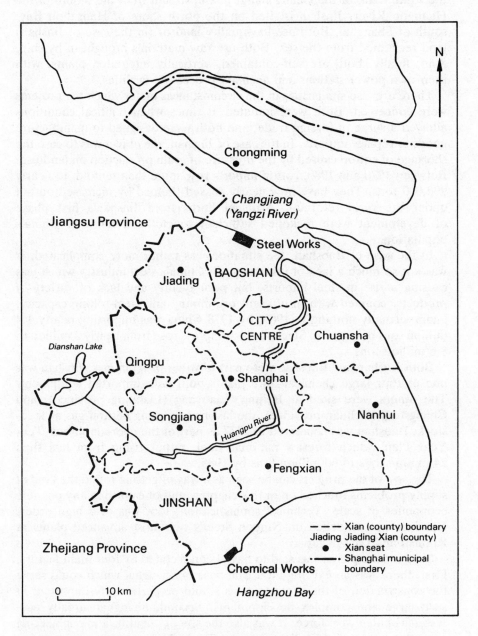

Figure 7.6 Shanghai: location of Jinshan and Baoshan
Source: Adapted from D.R. Phillips and A.G.O. Yeh (1987) *New Towns in East and Southeast Asia*, Hong Kong: Oxford University Press.

factors about the plants. Both are located on the coast. Baoshan is sited on the south bank of the Chang Jiang, just upstream from the mouth of the Huangpu River. Jinshan is sited on the north shore of Hangzhou Bay, south of Shanghai. Both use low-quality land or (in the case of Jinshan) land reclaimed from the sea. Both use raw materials brought in by ship. And, finally, both are self-contained, vertically integrated plants, with their own power stations and ancillary processing facilities.

There are also similarities in the circumstances under which the projects were conceived. Both were initiated at times when political conditions allowed a surge in foreign trade, and both were designed to manufacture goods to replace imports. In the case of Jinshan, the object was to ease the shortage of cotton caused by the pressure of grain production on land-use. Between 1970 and 1980, cotton imports rose more than tenfold, to nearly 900,000 tons. They have now nearly ceased (helped by increased output under the rural reforms). Artificial fibres from Jinshan's first phase of develpment were reckoned sufficient to clothe a tenth of China's population.

In the case of Baoshan, the situation was rather more complicated. It was not so much a lack of investment in China's steel industry which was causing a rise in steel imports, but poor quality and lack of variety of products, coupled with insufficient ore-mining and steel-rolling capacity. From virtually nothing in 1969, by 1978 China was importing nearly 1.4 million tons of pig-iron. Steel product imports rose from nearly 2 million to 8.6 million tons.

Both Jinshan and Baoshan were part of wider programmes. Jinshan was one of four large chemical fibre plants built with imported equipment. The others were sited at Beijing, Liaoyang (Liaoning province), and Changshou (Sichuan province), the last-named using natural gas as feed-stock. Baoshan was conceived in 1977 as part of the over-ambitious 'Ten-Year Plan' which foresaw national steel output rising from less than 24 million tons to 60 million tons by 1985.

The size of the projects can be seen as a way of getting round the kind of supply problems touched on in this chapter, and of capturing any possible economies of scale. Technical sophistication, too, was of a high order. Baoshan was modelled on Nippon Steel's two most advanced plants at Kimitsu and Oita in Japan.

Siting of the plants was said to have been dictated by four main factors. First, there was an existing industrial base in Shanghai which could serve the construction of the projects. This would be particularly important in such large and complex developments. Second, Shanghai already possessed a skilled workforce. It was also the site of scientific and engineering expertise, at the Fudan, Jiaotong, and Tongji Universities.

Third, Shanghai was the single biggest market for the projects' products, with more than 10 per cent of the national engineering industry and 13 per cent of textile output, and at the centre of the most important consuming region. And, fourth, coastal sites were considered to be most attractive for large industrial projects from the transport point of view. Since the railway network, particularly in east China, had reached saturation, it was not considered feasible to add large new flows of iron ore or oil to the system. Baoshan's riverside location enabled 97 per cent of raw materials – much of them iron ore from Australia and Brazil – and 70 per cent of products to be carried by ship. Jinshan has facilities for 25,000-ton ships to bring in oil from northeast China. Baoshan's iron ore, too, could have come from this source if enough investment had been made in mining capacity and land transport facilities.

The scale and siting of these projects may have been misconceived. Baoshan suffered from land sinkage and displacement in its foundation piles. The cost was grossly underestimated, and the project was suspended during 1980. It was a factor in the downfall of the premier, Hua Guofeng, who had backed it. Yet, in August 1981, the decision was made to continue it. Similarly, when sites were chosen in 1984 for the Seventh Plan's four ethylene projects, Jinshan featured among them.

Through these projects, we can see ways in which constraints on land, transport, foreign exchange and energy resources have combined with the enduring effects of past organisational and investment policies to shape the pattern of industry.

Coastal-based processing industries, insulated from the inland economy, importing raw materials and exporting final products, are now not only accepted but actively promoted. Despite contrary policies, whether egalitarian or market-based, industrial concentration and gigantism have continued to exercise a strong influence on the course of Chinese development. And Shanghai has been the biggest single beneficiary.

Case Study 7.2

SMALL-SCALE HYDRO ELECTRIC POWER

Eugene Chang*

Small hydro power (SHP) is like a good beer – it needs a good head, provides refreshing energy in times of need and reaches the parts that other energy sources cannot. For thirty five years or so, China has been promoting decentralised energy development for the large and scattered rural population. From modest beginnings, SHP development in China has gone from stength to strength. There is much room for future growth, since only 12 per cent of exploitable potential has been developed so far.

Far from being a fringe activity or passing fad, SHP in China is the lifeblood of industrial, agricultural, social and domestic life in many rural areas. SHP is defined as any hydro station whose installed capacity is less than 12,000 kW and with turbine unit capacity less than 6,000 kW (a typical one-bar electric heater needs 1 kW).

The layout for the most common type of SHP, a run-of-river type, is shown in Figure 7.7. A low diversion dam (1) is built across the river to divert water from the river to the head canal (2). A simple slab gate (3) at the headrace entrance is closed when maintenance is required. The headrace terminates in the forebay (4), which is simply a small settling basin. A side spillway (5) discharges any excess water, thus keeping the water level constant. Again, a gate (6) at the forebay entrance allows emptying of the forebay for maintenance.

Water flows from the forebay to the powerhouse (8) through a pipe called the penstock (7). Inside the powerhouse are the turbine and generator which produces electricity. Transmission lines then convey this electricity to the consumer. After passing through the turbine, the water flows back into the river.

The turbine consists of a runner which is rotated by the action of the impinging water. The rotation, and hence power (kilowatts produced) varies with the increase in the product of the height through which the water falls (from the forebay to the turbine) and the water flow-rate. The motion of the turbine is conveyed to the generator, which then produces electricity, just like a bicycle dynamo.

In almost every province and region, and especially in the ten southern provinces where rainfall is particularly abundant, SHP stations are to be found, with their turbines spinning away to feed the energy needs of the

---

* Eugene Chang is a consultant who has worked a number of times for the United Nations in China as an advisor on SHP.

Figure 7.7 Depiction of a typical small-scale hydro-electric power system

rural population. Priority is given to supplying productive end-users such as irrigation and drainage, processing of agricultural produce and rural small-scale industry. Lighting and other domestic uses are rather minor end-users.

Rural energy needs could, in part, be met by other sources. However, no other single renewable energy source has such versatility, or uses such a proven technology. Diesel generation, which is still the alternative in many countries, would not be feasible in China because of scarcity of fuel. Small-scale coal-fired plants are an interesting alternative, and would complement the seasonal nature of SHP quite well. However, this method is obviously restricted to areas with coal supplies. Bio-gas, with 5 million digesters in use, mainly in Sichuan, is largely restricted to domestic use. Windpower can only supply limited amounts of electricity, whilst solar power generation is not an economically competitive alternative.

Before 1950, there were only twenty-six or so SHP stations with a capacity of 2,000 kW. Then, during the 1950s, the national programme for agricultural development was formulated, and recommended that 'all water resources projects that can be utilised for generating electricity must be built to incorporate medium or small hydro power stations'. Thus, as part of national development policy, many SHP stations have been built, springing up like bamboo shoots after a spring rain, as the Chinese saying goes.

And they are springing up in all shapes and sizes. The smallest SHP stations do not actually need a powerhouse, and use specially designed micro-packages consisting of a turbine, generator and voltage regulator, all combined in a convenient portable package. A Nanjing organisation has developed a 650-KW model, just enough for a rice cooker. The cost is about 750 yuan, about half the price of a colour TV. At the other end of the SHP range, the stations begin to look like conventional hydro stations, with a dam and a reservoir, and can serve as backbone stations supplying electricity for almost a whole county. The station utilising the highest head (water drop) in China, and possibly in the world, is being constructed in Hubei province, using water falling through a vertical height of 830 metres.

Now there are over 74,000 SHP stations in operation in China, with installed capacity over 9,060 mW, annually generating over 20 billion kWh. This is about three times the 1984 capacity of Gezhouba Dam on the Chang Jiang, China's largest hydro station. SHP accounts for about one-quarter of total installed hydro capacity, and supplies about one-third of rural electricity demand.

Demand is rising. In the words of the minister responsible: 'At the beginning of the 1980s, the rural areas in China underwent a profound change from traditional susbsitence farming to modern agriculture. At this historic turning point, there is now an urgent need for electric power in the rural areas of the country.' New demands on SHP mean new demands on technical standards, especially at the planning stage, on levels of automation, on station efficiency and management and on profitable operation, especially as funding gets tighter. Increased international co-operation is also necessary for modernisation of China's SHP technology and policies, though China's wide experience has been recognised internationally for some time.

FURTHER READING

Further information on China's SHP can be found in:
World Water (1984), June: 31–3.
Water Power and Dam Construction 1985, February: special issue.
H.R.C. *Small Hydro Power in China: A Survey*, Intermediate Technology Publications Ltd.

Chapter eight

Urbanisation: processes, policies and patterns

Wing-Shing Tang and Alan Jenkins

Almost as many people live in towns and cities in China as in the whole of the United States. A best estimate of China's urban population at the end of 1988 was 230 million, about 21 per cent of the total. Later on we will have to clarify what is meant by 'urban' in China, but for the moment it is important to recognise this urban dimension of the country. It is one which is easily missed among the images of peasants working in the fields which more readily come to mind, given that many outsiders see China as an essentially rural society.

There is truth in this impression; at the beginning of the economic reform period in 1978, more than 85 per cent of the people lived in the countryside. But the urban image is important too: that of city inhabitants coping with cramped and poor housing, inadequate transport systems, high levels of air and water pollution, now all intensified by rising rates of rural–urban migration. China's geography is in prospect of being transformed by what may be the largest-scale urbanisation in human history.

THE URBAN HERITAGE

Cities of considerable size, and a widespread network of market and administrative towns, are not new in China. In the thirteenth century Marco Polo travelled from the commercially vibrant states of North Italy, and found in China a level of urbanisation and urban sophistication that astounded him. But later, as Europe industrialised and urbanisation increased, China's towns and cities remained islands of administration and commerce surrounded by a sea of rural life, with no similar transformation.

It is only really in the last forty years that an urbanisation based on industrialisation has emerged, largely concentrated in relatively few centres as a result of state-sponsored projects. But in the ten years of the reforms, partial but significant increases in the freedom of people to move around, and the commercial growth of parts of rural China, has led to new reasons for urban growth, including that of many small towns as well as the big cities. With this has come a lively debate in the country about the best path to follow, whether large-scale cities or many new or enlarged small and medium-sized towns.

New pressures for urban growth

Whatever the outcome, many people are voting with their feet and moving to town and city when possible. This new urban growth is in sharp contrast with the official emphasis of the pre-reform period, especially of the Cultural Revolution (1966–76), which was on rural achievements.

Agriculture and rural industrialisation were the state's priorities, and this struck a chord with many outside observers who were attracted to the idea of China modernising, industrialising, but not urbanising. Maoist China was thought to be quite different from other Third World countries, where development policies were deemed to favour urban elites, with impoverished peasants leaving rural areas and congregating in shanty towns on the fringes of the cities.

Despite this outsider view of the country modernising without the familiar horrors of urbanisation, China did actually experience large-scale urbanisation in the period 1949–60. And although urban growth was limited in the years 1960–76, many analysts now see this as part of policies which actually held back the development of rural areas, despite the rhetoric. Moreover, while urban growth was slowed, the needs of existing urban areas were severely neglected during the Cultural Revolution. Housing and infrastructure (transport, power, water and sewage) received very little investement, in many cases not even sufficient to keep them in good repair, let alone improve them. After the reform programme began in 1978, the level of investment in urban construction and infrastructure was much increased.

One factor crucial to the reforms has been the argument that the number of people needed in the countryside for agricultural production is much less than that under the commune system, possibly as much as a third of the rural population total (see Chapter six). It is this which makes the debate about how urban growth should occur – small towns or large cities – so significant. Before we can consider this problem and the impact of the reform period more generally, the events and policies of the period 1949–76 need to be examined to see how they shaped contemporary urbanisation.

The inherited pattern in 1949

Urbanisation can be defined simply as the process of increasing the concentration of a population into towns and cities (urban places), usually involving movements of people from rural areas in which agriculture and related activities are the main source of livelihood. There is a danger that 'urbanisation' is seen as a universal process, a concept transferable between different time periods and social and political systems without having much variation in its meaning. We need to acknowledge the very different reasons why urbanisation might be happening in different places, perhaps even within one country. This definition also masks the problem of what is an 'urban place', a problem we will come back to.

When the Communist Party of China (CPC) moved from its mainly rural guerilla base areas to enter the cities in 1949, it inherited a complex urban pattern. This included elements of the traditional pattern (of administrative and commercial towns and cities) on which various imperialist powers had superimposed some modern, Western-style urbanisation, mainly in coastal ports. Traditional urban places were generally walled settlements, housing government functionaries, merchants and related traders and artisans. Smaller market towns served the needs of exchange of produce between people in the surrounding rural areas.

The Western-influenced cities were often new urban centres (like Shanghai) based on international trade, with some manufacturing and financial services. They had attracted and retained many Chinese inhabitants and become some of the country's largest cities by the time the CPC came to power. These 'Treaty Ports' (see Chapter three) provided a significant legacy of industrial and infrastructural development for the new government to take over.

URBANISATION IN MAOIST CHINA

The central feature of urbanisation in China is that the process has been controlled to a large extent by the state and its bureacracy. This was made possible by policies adopted soon after 1949, which included the nationalisation of urban land and larger industries. The state or its local agents (i.e. city administrations) also took control of urban housing, along with larger retail and service establishments. Party and government units increasingly intervened to control private, small-scale urban commerce and services, including those in the complex web that tied together town and countryside. Even more significant were the rationing systems for essential goods, the permit system for place of residence, and general restrictions on individual travel which meant that ordinary movement around the country was very difficult without permission from the local authority.

From Liberation to the Great Leap Forward

The CPC's attitude, in the revolution and its aftermath, to cities and urbanisation seems to have been equivocal. On the one hand, urban centres and the educated urban classes had to be accepted and recognised, as they were essential for the administration of the state. But on the other, the educated urban classes were distrusted and seen as hostile to the regime because of their class position and foreign connections. The cities were seen as being parasitic and 'non-productive' because of these classes and the disdain with which urban commercial and imperialist activities were held. Places like Shanghai and Guangzhou (Canton) needed to be transformed from 'consumer' into 'producer' cities. As a result, investment was targeted on industrial production, while spending on commerce, housing and services was restricted.

Industrialisation without urbanisation was difficult, especially as the chosen method of growth in the economy in the 1950s was Soviet-style central planning, with investment concentrated in urban centres. This led to some conflict between the economic objective of rapid industrial growth and the political one of restrictions on urbanisation aimed at reducing the 'bourgeois' role of the cities. On the one hand, state planning bureaux supplied large numbers of workers to the new urban enterprises, arranged supplies of food and other basic goods, and built housing, schools and clinics. On the other, from 1953 onwards, the state introduced stringent controls on the movement of people, especially from countryside to city, by means of household registration regulations. Without being registered in a particular place a person could not live there. This was reinforced by the ration system of grain and cotton cloth, access to which was only possible through shops at an individual's registered location.

One consequence of the industrial development of the First Five-Year Plan (FYP) of 1953–7 was the industrial expansion of long-established administrative cities such as Xian and Nanjing and the creation of entirely new cities like Baotau, the iron and steel centre in Inner Mongolia. This also meant the emergence of industrial cities of a new type, away from the coast, whose location was based on very different considerations from those which influenced the modern development of the coastal cities. One factor in their development was military strategy: the need to place vulnerable enterprises at some distance from anticipated coastal invasion (see Chapters two and seven).

The First FYP was followed by the Great Leap Forward, in many ways a Maoist negation of the heavy industry approach of the Soviet-inspired plan, aimed at a different style of (rural) industrialisation. The rural problems which accompanied it, largely a result of the new economic arrangements of the commune system introduced from 1958, led to a large-scale influx of rural people into towns and cities. This was at its most intense during the famine disaster (1959–61), and it is estimated that during the Great Leap period about 16 million people moved to urban centres.

Considering the period 1949–60 as a whole, the urban population grew from 49 to 109 million, representing an increase in its share of China's total from about 9 per cent to more than 16 per cent (see Table 8.1). As a major authority on urban China put it: 'China's pace and scale of urbanisation during the first decade of the People's Republic is without parallel' (Kirkby 1985: 112).

MAOIST ANTI-URBANISM?

An abiding impression among outside observers is that under Mao Zedong's leadership China pursued an anti-urbanisation policy. Certainly, the role of the People's Communes as the basis of rural living, involving both agricultural development and small-scale industrialisation, dominated our impressions of how and where rural people were supposed to live. They were clearly not meant to move to towns and cities.

As a result, many outsiders considered that Maoist policy between 1966 and 1976 was meeting the people's needs without causing the unplanned and poverty-driven urban growth resulting from rural–urban migration which they found dispiriting in many other Third World countries.

But were Mao's policies really based on a concern to avoid the negative results of rapid urbanisation, or was this wishful thinking? In fact, it is much more likely that China's leaders had other reasons for controlling the growth of cities. And, today, outsiders are much more aware of the price in terms of the intimidating role of the state's power over people's lives involved in the prevention of movement and migration. In terms of the policy's impact on economic development, outsiders have gradually become aware that prevention of mobility created a 'great wall', effectively sealing off rural China from the benefits of commerce and migration.

The Great Leap and its aftermath

During the First FYP concern emerged as to how a growing urban population would be fed: if people moved to cities they could not grow grain, and those left in the countryside would have to produce more, which could then be procured by the state to sell to workers. The CPC'S concern for feeding the cities is linked with worries about its government's political security and urban unrest, a factor which has deep roots in Chinese history. This concern is reflected in the obsession with grain production which for so long dominated agricultural policy.

After the disaster of the Great Leap Forward and its resultant famine, many millions of peasant people moved to towns and cities to improve their survival chances, because of the grain storage and distribution systems and earning opportunities available there. This presented a severe challenge to the workings of the urban economy and reduced the number of agricultural producers available to grow subsequent crops. In the early 1960s, as many as 20 million peasant people were dislodged from urban areas and returned to the countryside.

The place-of-residence registration system

After this time it appears that the mobility of people was much more strictly controlled, using the measures set up in the mid-1950s, and that this situation remained until the reform period from 1978. The system was based on a rigid division of the population into two basic categories: either agricultural or non-agricultural households. All individuals living in urban areas had to have a registration booklet which specified where they were allowed to live. It also defined them as agricultural or non-agricultural, since urban areas normally include some farming land within the local administration's boundary.

People with an urban, non-agricultural registration were entitled to ration books for key goods (mainly grain, cooking oil and cotton products), were allocated jobs through the state employment bureaux, and access to schooling and health services either through their employer's own facilities or those of the urban authorities. This combination of registration and the lack of access to basic goods without the ration coupons (there was virtually no free market) very effectively prevented movement into and even between urban areas. It was policed by both the public security department and local neighbourhood committees who reported on movements of people to their areas.

De-urbanisation

From about 1960 then, state power was used very effectively to limit the growth of urban areas, though it appears that city authorities acting independently would arrange the recruitment of short-term supplies of labour from nearby rural areas. Movement was only permitted as part of national policy in circumstances when it suited state policy to increase recruitment to urban industries. One such time was the period of the *sanxian* policy from 1965 to 1972, when secret military and industrial development in the southwest led to new cities and the expansion of existing ones like Chengdu (see Chapters two and seven). In general, though, migration from the countryside to the city was strictly controlled, and permission to live in urban areas seldom granted, except for limited periods.

But not only was there restriction on flows into urban areas; there was from time to time active encouragement or compulsion for people to move out of the cities and towns and take up new lives in the countryside or small towns. Many moves were the result of political mass campaigns, in which many went 'down to the countryside', sometimes to be educated by the rural masses in 'correct' political thinking. One major campaign, during the Cultural Revolution, possibly involved more than 20 million 'educated youth' going down to the countryside supposedly to pass on their knowledge (e.g. literacy) and more modern ways. The compulsion in-volved, and the lack of freedom to move back to their homes, made many ill-disposed to the success of the campaign.

More than this compulsion, there were also exile campaigns aimed at 're-educating' sections of the urban population who were considered politically dubious. The idea behind it was that being 'red' was more important than being 'expert' (see Chapter one). Such exile was much more clearly a form of punishment. It could be open-ended or temporary, depending on a person's status and the nature of the campaign at a particular time. That political campaigns, government encouragement or outright compulsion have been significant in the movement of tens of millions of Chinese around the country further underlines how significant has been the role of the state in determining urban patterns.

THE PROBLEM OF DEFINING 'URBAN'

One's understanding of the complexities of Chinese urbanisation is not helped by the difficulties in using Chinese data. In particular, it has been extremely difficult to comprehend what exactly is meant by the term 'urban' in China. Definitions have shifted, and it also seems that different organisations within the country have themselves used varying definitions. Does the term 'urban' involve the size of a settlement or the density of population in a place or area? Is it rather a matter of the economic characteristics of the people in a particular place or area, the sort of work they do and what proportion of their livelihood is derived from things other than farming? How is the extent of a place or area determined when it is to be designated 'urban'? What about households in which the various members have different types of job?

It is especially difficult if the definition of 'urban' changes over a period of time, as it has in China, so that one year's figures cannot really be compared with another. There is also the problem of making comparisons between countries, not many of which adopt the same definition of the word 'urban'. Also in China, the territorial extents of administrative units of local government have been altered in many places, so that rural areas surrounding towns and cities have been incorporated by urban authorities into their area. At a stroke this may increase significantly the supposed urban population, and great care has to be taken to ensure that such errors are avoided. One current official figure for the total urban population of China for 1988 is 550 million, which is patently absurd as it accounts for more than half the people of the country (Table 8.1). Some explanation is needed as to why such misleading figures are used in official Chinese publications.

One major reason for these high figures is the inclusion of agricultural households in data for urban areas. This has been even further exaggerated

Table 8.1 Urbanisation: urban and total population and a comparison of 'official' Chinese figures with modified series, 1949–88

Year	Official figures given since 1982 census[a]		Acceptable modified series[b]		Total population[a]
	Urban population	Per cent of total	Urban population	Per cent of total	
1949	57.65	10.6	49.00	9.1	541.17
1953	78.26	13.3	64.64	11.0	587.96
1957	99.49	15.4	82.18	12.7	646.53
1960	130.73	19.7	109.55	16.5	662.07
1964	129.50	18.4	98.85	14.0	704.99
1969	141.17	17.5	100.65	12.5	806.71
1976	163.41	17.4	113.42	12.1	937.17
1978	172.45	17.9	122.78	12.8	962.59
1980	191.40	19.4	140.28	14.2	987.05
1982	211.31	20.8	152.91	15.1	1,015.90
1984	331.36	31.9	174.42	16.8	1,038.76
1986	441.03	41.4	200.90	18.9	1,065.29
1988	550.00	50.0	230.05	20.9	1,100.00

Sources: [a] State Statistical Bureau (1988) *Zhongguo Tongji Nianjian 1988 (Statistical Yearbook of China 1988)*, Beijing: China Statistical Publishing House, p. 97. The figure for 1988 is an estimate.
[b] 1949–76 figures based on L.J.C. Ma and Cui Gonghao (1987) 'Administrative Changes and Urban Population in China', *Annals of the Association of American Geographers*, 77,3: 373–95, Table 7. 1976–82 from Kirkby (1982) 1984–88 estimates from an unpublished paper by Kirkby (1989)
Note: The main reason that recent official figures give such large urban populations are: (1) the municipal data which make up the figures is inflated by the inclusion of large numbers of rural people who happen to reside in the suburban city districts of municipalities, especially when the boundaries of such districts are arbitrarily drawn to include large areas of countryside; (2) the town data incorporated in the grand total includes vast numbers of agricultural people. (Population in millions.)

because in the 1980s many settlements were upgraded to the status of designated town and city. For instance, in 1975 there were just 194 designated municipalities (cities), and 2,800 towns. By the end of 1987, these had increased to 381 and about 8,000 respectively. This statistical increase in the number of urban areas does indicate something of the real nature of increase in the concentration of population, especially at the lower end of the size range. But the official figures for urban population within towns and cities are very misleading. For instance, the number of designated urban places was greatly increased between 1983 and 1984, suggesting a jump from 241 million to 330 million urban inhabitants. But the increase reflects the fact that many municipalities were annexing neighbouring counties, thereby absorbing many more agricultural households into their supposedly urban totals.

Estimates of the urban population

One way to 'define' urban is to count the number of people in households which are considered as non-agricultural who are living in designated cities and towns. On this basis, about 175 million people counted as urban in

1985, or about 17 per cent of the national total. The very high figure of more than half of the country's total population is derived from a much wider definition which includes agricultural households residing in urban places. Given the penchant in the last decade for urban authorities to annexe surrounding counties into the urban administration, it is no surprise that such large but meaningless figures are given.

One of the main uses of knowing the proportion of urban to total population (and the rate at which it is changing) is to help understand the shift in employment from agriculture to non-rural livelihoods. If this distinction is muddled, then urbanisation becomes useless as a concept for understanding change in Chinese society. As two Chinese officials commented recently, 'incorporating the rural people into the city population is improper and will contribute to misjudgements of the level of urbanisation in China' (*Beijing Review* 1989, 22–28 May: 22). Depending on the definition used, very different sets of figures for the urban population have been compiled in China and by foreign researchers. Now that we are aware of the difficulties involved, a wary approach can be made to such data. Without such care, very wrong interpretations of the level and speed of urbanisation may be made. Table 8.1 shows a comparison between one set of official Chinese statistics, and those which have been compiled from a combination of sources, which we believe represent a better view of urban population.

Growth and decline in urban areas

Our own figures clearly show the rapid rise in urban population in the 1950s, during the first phase of industrialisation, and the very large increase for 1960, indicating the rural influx generated by the famine. Conversely, the 1960s show an absolute decline in the population of urban areas as the peasants returned and then in the Cultural Revolution people were sent down to the countryside. The 1970s show a remarkable stability in the proportion of urban dwellers until 1980, when the impact of the economic reforms begins to show, with a rapid rise in the number.

By 1988, the decade of reforms had been accompanied by almost a doubling of the urban population. Much of the increase results from the ability of people without permits to live in cities and buy necessities on the much expanded free market. Many more have contributed to the growth of small rural towns, which have grown in many areas with the commercial and industrial expansion which has accompanied the rural reforms. These recent processes will be considered later.

The rise in absolute numbers of people in the towns and cities in the late 1960s and 1970s was not matched by sufficient investment in housing and infrastructure. The Maoist period left a poor legacy to Deng's China, with

211

years of inadequate urban planning as well. There was limited co-ordination between growth in employment and the transport, housing and other facilities which needed to grow with it.

This imbalance was partly due to the emphasis on heavy industry and high rates of capital accumulation to pay for it. Less of the national 'cake' was left available for consumption, including housing and infrastructure. Despite the rhetoric about the primacy of rural development during Mao's leadership, the state-run economy was dominated by a desire for increased heavy industrialisation, with the people being told that sacrifices in consumption would result in greater gains later on. In fact, taking the average floor-space in housing available to urban inhabitants, arguably one of the major aspects of the quality of their life, the per capita figure declined until the reforms began and shifted economic policy. In general, the larger the city, the worse the overcrowding and relative lack of investment in infrastructure.

REDIRECTIONS OF URBAN POLICY AFTER 1976

After Mao Zedong's death in 1976, and especially since the reform programme began after 1978, policies affecting both urbanisation and the internal development of towns and cities changed considerably. The industrial and rural reform policies had considerable effect on urban growth and patterns of urbanisation, as also did regional priorities. Not only has urban growth been much more rapid under the reforms, the size and type of urban place affected has been different from in the past.

Some of the impact on urban areas has resulted from the change in overall economic policy, linked with active attempts to influence their development through the shift in investement towards 'consumption'. The previous attitude that 'production must come first and living conditions later' was questioned. A backlog of investment missed in previous decades was recognised as needing action. An academic wrote in 1982 that urban dwellers needed 'electricity, heating, grain, clothing, transport, culture, recreation, health, and above all housing' (cited in Kirkby 1985: 169).

Concern for housing

State investment in urban housing has increased significantly since 1979. (In rural areas, where housing has always been privately owned, a boom in new house-building has occurred under the impact of rising incomes and the reduced risk of political criticism for 'bourgeois living'). The urban housing expansion has been under the direction of municipal housing bureaux and also that of the state industrial enterprises which have responsibility for their own workers' housing. The construction boom has become a very evident feature of the urban landscape, though there is also renovation of existing housing units. In general, since 1979, nearly three times as much housing space has been built each year as in the 1970s.

These policies have begun a considerable improvement in urban living. There are new regulations for housing standards, which include households not having to share kitchens. Other measures affect sewage services, pollution control, transport, domestic water and gas supplies. In a few cities, loans from international agencies and foreign investment have been used to improve infrastructure. In all, though, the level of state and local authority investment has still not been sufficient to make up for the years of urban neglect.

The inhabitants have seen some improvement in their facilities in other ways, through the growth of private enterprise and markets in services and goods which had been very difficult to get during the Cultural Revolution. The Chinese city, for so long a dull and starkly regulated scene, has taken on more of the traditional vibrancy, with street traders and artisans seeking to supply peoples' needs. Newly available supplies of vegetables, clothes, cooked food, restaurants, and services like bicycle repairers, have brought significant improvements to urban lifestyles. The new businesses are often set up by people who were previously expelled to rural areas and who returned to their home cities under the more liberal atmosphere of the reforms. This has helped them, and has also helped the authorities by avoiding unemployment and a social crisis.

Housing and investment conflicts

Although the housing stock has increased considerably, and standards have improved, new pressures for urban growth have meant that more people have to be catered for. Moreover, although the reform period has seen a new commitment to increased investment in consumption (including housing and urban infrastructure), it has also given priority to light industry and certain heavy-industry projects in cities. So there has remained a degree of competition for investment funds, which urban improvement has tended to lose.

The remedy proposed resembles the 'market' methods which have been intended as solutions to other Chinese problems. In 1980, the State Council (in effect the Cabinet) decreed that urban housing should gradually be 'commercialised'. This move towards a market economy for housing was based on the argument that the very low rents charged to urban tenants should no longer be seen as a welfare benefit. The rents were insufficient to generate the capital needed for reinvestment, and so reduced the welfare of other people for whom new housing could not be built.

There were even announcements about the need to reduce the standards of new housing units, because a reduction in unit costs would mean that more units in total could be built. In addition, considerable discussion and some experimentation was carried out on the sale of urban housing to tenants.

These policies to 'commercialise' housing have been only partially implemented, and raise major ideological problems as to how far a 'socialist' state can encourage a commodity economy in such things as urban real estate. The policies also fuel inflation, and there is evidence that they have considerably reduced urban support for the CPC's reform policies. Many urban dwellers, particularly those on fixed incomes from government jobs, have suffered from high rates of inflation in the last half of the 1980s and have not been impressed by the impact on them of the Dengist reforms.

New pressures in urban growth

Since the beginning of the reforms in 1978, urban population has grown substantially both in total and as a proportion of the national total (Table 8.1). Although it is in towns and cities that population control and the one-child policy have been most effective, the urban birth rate has risen during the 1980s. This is due largely to the age structure of the population, which includes a large share of people born in the baby-boom of the early 1960s v'.ə are now couples of child-raising age.

But we need to look for other causes of the increase as well, since it cannot be explained solely by natural increase. In 1978, the urban proportion of the country's total population was, we estimate, about 12.8 per cent. By 1988 it had risen to nearly 21 per cent, representing almost a doubling in the number of urban inhabitants to 230 million.

Part of the growth is explained by the fact that the government has allowed many of those sent to the countryside in earlier political campaigns to return to the cities. Since most of these are also in the child-rearing age-group, this portion of the urban population has been made even greater. But the most important factor contributiong to urban growth in this period is more general rural–urban migration.

After years of strict controls, the 1980s have witnessed tacit state approval of such movement, even to the larger cities. The factors which 'pull' people to the cities and towns are mainly the availability of new forms of waged employment or individual enterprise, though there seems also to be considerable speculative migration with no certainty of immediate employment. State policy seems to be to allow this, provided that in-comers do not rely on state rations; many, though, are able to register only as temporary workers.

Urban areas are attractive because opportunities in rural China have diminished and livelihoods are less secure with the break-up of the collectives. These 'push' factors are officially recognised in the government's estimates that in the mid-1980s around one-third of the rural population was surplus to the needs of agricultural production. Some of the labour not needed in crop production has been encouraged into other activities in rural areas, including forestry and, especially, new private and co-operative enterprises. But much is already 'footloose' and contributing to the growth of small towns and to a marginalised informal sector in cities (see Case Study 5.1 on p. 126). Recent estimates suggest there is a 'mobile population' amounting already to about 70 million.

Changes in the registration system

Though the household registration system remains in force, it has been significantly modified. Indeed the regulations often seem to ratify what has already taken place. In 1983 the State Council permitted the hiring of labour in the countryside, and this signalled acceptance of the emergence of a market for workers. Then, and also in 1985, it enabled rural labourers 'to leave the land but not the countryside'; provided their families could supply them with grain, people could take up work in the smaller towns.

The use of 'temporary residence certificates' has increased, to supply urban industry with rural labour, usually male, on short-term contracts (often for six months). This has resulted in some rural areas being largely populated by women, children and the elderly. Estimates suggest that such temporary residents can account for up to 10 per cent of city inhabitants. In 1985 a survey of Shanghai indicated around 1.6 million visitors there for all sorts of reasons, but many were certainly intending to be in the municipality long term. What is uncertain are the numbers of illegal migrants. The expansion of the private sector of the economy has meant that people can find work, food and clothing without recourse to the state system, though housing is much more difficult.

These changes reflect a significant re-evaluation of the relations between city and countryside, and an end to the barrier between the two in terms of population movement. With the change to the responsibility system in agriculture, and the increase in commercial operations in both cities and smaller towns, connections with the countryside have become much more direct. Even state-owned urban industries are being encouraged to develop stronger links with enterprises in adjacent rural areas (see Case Study 8.1 on p. 220). A pattern has developed where the cities concentrate on higher-level technology while the rural entrerprises linked to them use lower-level technology and make their contribution through a greater (and cheaper) labour input.

It is important to emphasise that such developments are spatially concentrated in rural areas nearer the larger and more prosperous cities. Together with the expansion of rural housing, such developments are eating away at China's limited arable land, and hard policy decisions will be needed if the inherent dangers of this are to be avoided. Much of this type of growth is in many dispersed small towns, and this has helped fuel the sharp division between state-planners and academics over whether urbanisation should instead be directed to larger cities which, their supporters argue, make less wasteful use of land and more efficient urban places.

URBAN FUTURES: LARGE, MEDIUM OR SMALL?

During the reform decade of Deng Xiaoping's leadership, and in contrast with the impression given under Mao, China's future has been clearly seen as urban. In 1982, an article in the Chinese journal *City Planning* stated views that would have been heretical a few years before: 'Urbanisation is a necessary consequence of the economic development of society, whatever the country, whatever the societal system, admitting absolutely no exception' (cited in Kirkby 1985: 221).

Debate over city growth

Increasingly, economic planners and academics highlight the economic benefits of production being concentrated in large-scale urban centres. Economies of scale, including the greater efficiency of investment in infrastructure when concentrated, are used to explain the advantages of a city-based strategy. There is still support for a different 'Chinese' path, but its advocates appear increasingly unsure of their position. Before 1976, many planners were concerned about the growth of the larger cities. For most of the period 1960 to 1978, official policy was to restrict the size of the largest ones. It is also probable that, as with regional policy, there were military reasons for not wanting to increase the concentration of people.

The debate about the urban preferences involve certain defined size categories. Cities with a population over a million are termed 'extra-large'; those between 500,000 and a million are 'large'; between 200,000 and 500,000 they are termed 'medium'; and those less than this are called 'small'. It is the extra-large and large categories which are usually considered as needing to be controlled.

At the start of the economic reform period it seemed that most articles in the Chinese press and academic journals were praising the development of the small-town strategy. In 1980, a national conference on urban planning heard the policy announced to 'strictly control the development of large cities, rationally develop medium-size cities, and vigorously promote the development of small cities and towns' (Kirkby 1985: 207). Small-scale industries in them were seen as a way of soaking up surplus rural labour. The increased amount of produce sold through the private sector suggested a revival of the traditional web binding towns at different levels of the urban hierarchy with their rural hinterlands. The role of the small market towns in the economy was given increased official recognition.

The tension between this view and that of the policy to promote larger cities soon emerged. By 1982, seventeen Key Point cities had been designated, all with more than a million population, such as Shanghai, Guangzhou, Tianjin, Dalian, Chongqing and Wuhan. They were given much greater planning and financial autonomy and allowed to retain more of the revenue earned by enterprises under their control. In effect, this made it more difficult for there to be much central control over their growth and pattern of urbanisation. The justification was the need to produce the maximum benefit for the national economy by permitting these cities the opportunity to concentrate scarce resources in a manner considered to be more efficient by some leaders.

As part of this acceptance of the role of extra-large city growth, also in the early 1980s there was a revival of interest in a policy of satellite towns (rather like British New Towns) for cities such as Beijing and Shanghai. This is an attempt to obtain the benefits of agglomeration while diverting urban development away from the most productive farmland. Such policies suggest that the larger categories of cities will, in fact, continue to grow, and will be less responsive to central direction.

An increased role for planning?

One of the paradoxes of the 1980s' urban reforms is that they have included an increased role for urban planning, while at the same time cautious moves have been made by the government to treat urban land as a leasehold commodity which should respond to market forces. As a part of this process, there has also been the suggestion that enterprises should be charged for their use of land. From the 1950s to the 1980s, because urban land had been nationalised, it was considered unnecessary to charge state-owned enterprises for its use. This had led to its wasteful use, and considerable difficulty in wresting it from an enterprise's control even when no longer needed.

Urban planning has had a mixed record in China. In the First FYP a corps of professional urban planners was formed, and most large cities drew up plans to guide industrial, residential and other development. The concern was mainly with physical structures and layout. After 1960 urban (and regional) planning was eclipsed, and the planners dispersed. Since 1978 a new framework for urban planning has been created and groups of planners reassembled. In particular, urban and regional planning seems to be emerging as a combined discipline.

The new cohort of planners are therefore involved in a very different sort of activity, requiring the development of new ideas about the urban land market, systems of land rent and land use taxation. This is an area of considerable difficulty in a socialist state, involving the challenging of conventional thinking which would prohibit land being treated as a commodity. It will remain an issue of contention between the reformers and the 'conservatives', who favour more centralised control of such aspects of the economy.

Whichever policy is adopted, the problems identified with the old system of land allocation, and lack of charging, will need to be dealt with. To facilitate rapid industrialisation in the past, land has been granted almost free of charge. This means it is wasted by low intensity of use, resulting in inefficient developments. The implications of any major changes in this system for the relative merits of the continued growth of the largest cities are interesting. Do the agglomeration and development based on economies of scale presumed to be available in large cities still apply if market charges are made for land? What are the implications of this for the pattern of urbanisation and the balance of development between urban areas of different sizes?

Answers to these questions will take some time to become clear, and the introduction of new systems which challenge the entrenched interests of existing enterprises and urban authorities will take time to be implemented. But a Land Management Law was passed in 1986 which provided the legal backing for more systematic land management. Since then, every major city has started to re-register user rights for its land users, and land leasehold maps are drawn up for management purposes.

Some experimental systems are being tried, as for example at Fushun in Liaoning province. There, a levy is imposed on the use of land, and this provides revenue for further urban development. Some land was also handed over to the city, presumably surplus to need and for which the holders did not want to pay charges. Other plots of land were exchanged between different land users, reducing the conflict between residential and production areas. These are features which are far from prevalent in China, but if the urban reforms continue in this direction there could be significant changes in urban morphology and urbanisation.

CONCLUSION

To some extent, what is seen to be happening in China's urbanisation in the 1980s resembles what happened in the 1950s, with large cities as centres of modernisation, movement of rural people to urban areas, a concern for urban planning, and emphasis on urban construction. The major difference is that small-town growth has been of tremendous importance, in parallel with the rural reforms and the rapidly growing commercial economy in the countryside.

Taking into account the problems with the data in knowing what is real urban growth, we consider that in 1978, at the beginning of the reform period, the urban population was 123 million and that by the end of 1988 this figure had increased to 230 million. Continued growth of this order will pose enormous problems for the development of adequate urban infrastructure, especially housing.

If rural industry maintains its rapid growth rates it may siphon off much of the surplus rural workforce into smaller cities and towns. But this would require some state control over migration and policies to allow rural enterprises to survive in a competitive market in which more technically advanced urban industries will be favoured. Control would certainly also need a co-ordinated urban and regional strategy, especially if the policy to favour the rapid development of the Coastal Region continues. The likelihood is that with decreased control over the movement of people, this development will continue and that China may experience one of the largest-scale urbanisations in human history.

ACKNOWLEDGEMENT

This chapter has benefited from the comments and editing of Richard Kirkby and Terry Cannon.

REFERENCES

Kirkby, R. J. R. (1985) *Urbanization in China: Town and Country in a Developing Economy 1949–2000 AD*, London: Croom Helm.

FURTHER READING

Chan, Kam and Xu, Xueqiang (1985) 'Urban population growth and urbanization in China since 1949: reconstructing baseline', *The China Quarterly* 104: 583–613.
Chan, Kam Wing (1988) 'Rural–urban migration in China, 1950-1982: estimates and analysis', *Urban Geography* 9(1): 53–84.
Kirkby, R.J.R. (1985) *Urbanization in China: Town and Country in a Developing Economy 1949–2000 AD*, London: Croom Helm.
An invaluable source, a comprehensive study of urbanisation and urban–rural relations.
Kojima, R. (1987) *Urbanization and Urban Problems in China*, Tokyo: Institute of Development Economics.
Useful essays on aspects of urban planning, including housing, pollution, transport and urban renovation.
Ma, Laurence J. C. and Cui Gonghao (1987), 'Administrative changes and urban population in China', *Annals of the Association of American Geographers* 77(3): 373–95.
Whyte, M. K. and Parish, W. L. (1984) *Urban Life in Contemporary China*, Chicago: University of Chicago Press.
A sociological study of many aspects of urban life, including rural–urban migration.
Zweig, D. (1987) 'From village to city: reforming urban–rural relations in China', *International Regional Science Review* 2(1): 43–58.

Case Study 8.1

WUXI: A NEW MAJOR URBAN CENTRE

Wing-Shing Tang

Wuxi is in the southern part of Jiangsu, to the northwest of Shanghai. While its name is unlikely to be known to many people outside China, this city is now the country's fifth largest in manufacturing, surpassed in terms of the value of output only by Shanghai, Beijing, Tianjin, and its neighbour Suzhou. Although it is therefore rather exceptional (in 1984 it was twelfth), the processes involved in its growth and that of the surrounding area warrants discussion as a case study. (This area, south Jiangsu province, is also discussed in Case Study 6.1 on p. 160 as an example of what has been happening around Nanjing.)

Since 1983, Wuxi city has administered three counties: Wuxi, Yixing and Jiangyin. To avoid confusion, these four administrative units may be grouped together and called unofficially the Wuxi region. Wuxi itself is a 'designated city' (DC), a category of urban settlement which confers certain local government status.

In 1979 Wuxi DC had a population of 587,700 non-agricultural people (NAP), with a comparatively dense network of economic activities in the city proper and many industrial nodes in the countryside. Situated in the Chang Jiang (Yangzi) River Delta 'fish and rice country', the rural economy has a strong tradition of cultivation, side-line production and handicraft industries running side by side. The industrialisation of the countryside happened here during the Cultural Revolution, and by 1977 total rural industrial output of Wuxi County surpassed that of the DC.

Like many other areas in China, Wuxi had a surplus of agricultural labourers. Many of these have been able to find employment in the dense system of towns in the area. Drawing on the comparatively rich agricultural resources in its vicinity, Wuxi DC has itself emerged as a regional economic centre, with the nickname of 'Little Shanghai'. By the mid-1970s, Wuxi DC had developed engineering and electronic industries, in addition to the more traditional textiles and food processing.

In addition to its light industry, it is important to note that the development of Wuxi was largely attributable to local capital re-investment by collective-owned enterprises rather than to the state. (Jiangsu was one of the provinces that after 1958 did not directly manage industrial enterprises, leaving most of them subordinated to the DC and county administrations.)

The shift in national policies to the economic reforms after 1978 was applauded by Jiangsu province and Wuxi DC authorities, because their strength already lay in the sorts of activity which the reforms were meant to promote. For Wuxi DC itself there was a consensus as to what the province and Wuxi wanted to develop: light industry and agriculture in Wuxi.

But in the late 1970s, shortages in jobs, land, raw materials, energy and consumer goods made the local situation difficult. Rural industrialisation became the panacea, and there was a drastic increase in the number of production centres in the countryside. But there was also much competition for land in the vicinity of the city proper.

To help avoid these conflicts, enterprises in the suburban counties became more appropriate, but new problems arose as these were under different jurisdictions. As a response to this problem, the province government decided (as elsewhere) that the other two counties should be administered by Wuxi DC (see Figure 6.4 on p. 154).

There were other developments in urban industries that added to rural industrialisation. In textiles, expansion led to product diversification. International competition, a result of turning Jiangsu into an export centre for textiles products, also changed the nature of the production process. Technical improvements required a textiles machinery and equipment industry. Organisational change was also needed to accommodate customers making orders at short notice, or smaller orders than usual. Since the existing enterprises, usually crowded in the city proper, were spatially unable to adjust to these new circumstances, one of the solutions was to integrate with other enterprises in the rural area.

Similar developments could be detected in engineering, Wuxi's second-largest industry in value terms. One of the major problems in developing new machines for light industry was the shortage of material resources, especially steel products. Being mostly collective-owned with meagre allocation, the enterprises had to secure the supply by integrating with suppliers. The most important concern of the electronics industry, Wuxi's third largest, was to raise the production of consumer electrical appliances for the insatiable market.

To capture a bigger market share, state-owned enterprises have increased their activities in outlying areas. But collective-owned enterprises in the counties, aware of the competition, follow suit. They have transferred part of the production process to rural households (like outworking), and have integrated with enterprises in other counties in exchange for energy and raw materials (for which the Wuxi region is so desperate). In other words, the counties in the Wuxi region in the mid-1980s developed to such an extent that they were not only the passive recipients of economic integration but also active initiators. The outcome is a boom in activities in the counties.

Another important factor for the Wuxi region is the influence of Shanghai. Some of the effects of industrial restructuring in Shanghai under the economic reforms have led to more development in the Wuxi region than would have been expected if they had not been so close to one another. This has been especially the case since the establishment of the Shanghai Economic Region in 1982.

Under the impact of the various factors, a much more complicated composition of the working population is emerging. Prior to the reform period, peasants were mainly farming, with a smaller proportion involved in industrial activities. Rapid rural industrialisation has drastically changed this. In 1979 in the suburban district of Wuxi DC alone, 36,800 peasants worked in rural industries (45 per cent of the rural labour-force). By 1986 this had jumped to 51,900, or 70 per cent of the rural labour-force. For the three counties as a whole, there were 298,000 in rural industry in 1979, and this has risen to 736,300 (46 per cent) in 1986. In short, after a few years of development, peasants who work in rural industries have become a significant component of the working population.

In comparison, the growth in numbers of state employees has been more modest. By 1986 the total for the DC increased to 501,600 – from 430,300 in 1983. Individual workers (mostly involved in personal service trade), almost unheard of in the early 1970s, re-emerged in the mid-1980s. The rate of growth between 1985 and 1986 was reasonable for the DC (9 per cent) and astronomical for the three counties (52 per cent).

The investment pattern reflects the dynamics of the collective-owned (i.e. local town and county) enterprises. On average, they made up 19 per cent of the total investment in fixed assests from 1983 to 1986 in Wuxi DC, and 50 per cent in the counties. Of non-productive investment, housing accounted for, on average, 65 per cent of the Wuxi DC total and 47 per cent of the counties' total. In other words, because collective-owned enterprises dominate the economy, both the DC and the counties have been able to invest more on housing and other non-productive items.

These investments have required a lot of construction activities, and between 1980 and 1985, 7.73 million sq. m of floor space were completed. This was equivalent to the total completion between 1961 and 1980. The 1980–5 figure for housing was 3.97 million sq. m, which was 1.5 times the total between 1949 and 1980. Some of the construction activities have encroached on arable land in the countryside. The 1985 figures of land used for capital construction and peasant house-building were 138.5 ha for Wuxi DC and 397.8 ha for the counties as a whole. One year later, the figures were 146.7 ha and 1,550.3 ha, respectively. These figures imply that, at least in recent years, the pace of land conversion has been persistent in the city, and has increased significantly in the counties.

The population of Wuxi DC (excluding Wuxi County) increased from 781,000 in 1981 to 860,800 in 1986. The 10.2 per cent increase is, surprisingly, much smaller than that of all the DCs in the nation (68 per cent). The non-agricultural population figures for Wuxi DC were 587,700 in 1979 and 724,600 in 1986. This is a 23.3 per cent increase, in comparison with the 41.2 per cent national figure, and this indicates the already high level of NAP in Wuxi at the beginning of the reforms. For the region, official data shows that total population increased from 3,870,100 in 1983 to 3,954,600 by the end of 1986, a very modest increase, which is actually lower than the assumed rate of natural increase. This poses a puzzle, still to be resolved, as to why the rates of increase for the city and region are so low.

Chapter nine

Foreign investment and trade: impact on spatial structure of the economy

David R. Phillips and Anthony Gar-On Yeh

Trade is important for China as it is for most countries. However, its significance to China has fluctuated considerably over the past 150 years, and especially since 1949. Between 1949 and 1978 the volume of trade was generally small, and the major trading partners were mainly communist countries, particularly during the 1950s and 1960s. 1978 and the adoption of the 'open door' policy marked a turning point in foreign trade and investment in China. Since then there has been a great increase in the volume of trade, especially with non-communist countries. Even so, in the mid-1980s, with about one-fifth of the world's population, China still accounted for only about 5 per cent of world trade: clearly even a relatively small rise in trading activity could have a significant impact on the pattern of world trade.

China has largely abandoned its earlier attitudes hostile to foreign investment. Today, it has recognised that mutual gains are available from foreign trade although the benefits may be inefficiently and unequally distributed: workplace and marketplace are currently insufficiently sensitive to allocate appropriate rewards to individuals and enterprises. Legal and administrative reforms have been introduced to attract foreign investment in conjunction with modernisation. In addition, Special Economic Zones, open coastal cities, and other areas have been designated for the purpose of attracting foreign investment. The 'open door' policy has had great impact on the pattern and volume of foreign trade and investment, and also on the spatial structure of China's economy, influencing the pace of development in different provinces and areas of China.

THE 'OPEN DOOR' POLICY AND MODERNISATION

At the important Communist Party meeting in December 1978, it was decided to focus the country's efforts not on politics and 'class struggle' but on economics and the Four Modernisations (see Chapter one). This involved reducing the dominance of central state planning and permitting the more market-led system of the 'open door' policy. This includes an 'open door' to world markets and foreign investments. Efforts to promote economic and technical co-operation with other countries were begun. The result is that China is no longer isolated from foreign contacts and instead actively promotes economic ties and technological exchanges with international bodies and other countries. Nevertheless, the government retains a desire to maintain economic independence and to avoid earlier inequalities in foreign relations.

The place of foreign trade in modernisation

By promoting foreign trade, China hoped to accumulate the funds and technology necessary for modernisation. Foreign trade has not always been important in the Chinese economy which before 1978 practised extreme self-reliance to avoid contamination by the capitalist world. For example, the annual volume of trade between 1950 and 1974 averaged no more than 4 per cent of gross national product (GNP). This has risen considerably since the 'open door' policy, to reach over 10 per cent of GNP. Between 1965 and 1980, the average annual growth rate of exports was 5.5 per cent and imports 8 per cent; between 1980 and 1985, this average had increased further to 8.8 per cent per annum for exports and 17.6 per cent for imports.

The resulting deficits in the balance of trade have caused severe problems and a drain on foreign currency and gold reserves. Instead of paying for technology by raising exports (which has been too slow), there has been a growing need to borrow from foreign governments, banks and international agencies like the World Bank. This delays the need for increasing the export capacity, but it also intensifies the pressures on the domestic economy to find new goods to sell abroad. This, too, has its regional impact through the emphasis put on crop exports and processing of manufactures from some areas.

Many of the reforms effected under Deng Xiaoping's leadership have been institutional, directly and indirectly aimed at promoting trade. At the same time the government has tried to maintain control of foreign earnings and payments, with mixed success. Because a highly centralised economy is often incapable of handling trade expansion efficiently, decentralisation of decision-making has been growing, perhaps at times getting out of hand, making the task of controlling the balance of trade even more difficult.

The recent role of foreign trade (particularly of imports) has been to facilitate and accelerate modernisation and economic development by enabling the acquisition of raw materials, machinery, equipment and technology which are not available or cannot be produced in sufficient quantities domestically. This shift, and the use of exports to attempt to pay for more imports, means the volume of trade has inevitably grown (Figure 9.1). Separate parts of the country will inevitably be differentially influenced by this expanded trade, and these spatial aspects of production and trade are discussed later.

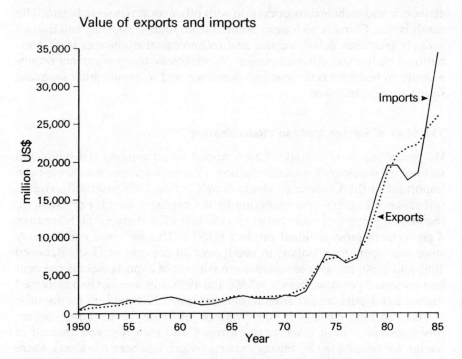

Figure 9.1 Graph showing exports and imports by value, 1950–85

Direction and control of trade

From 1952 to 1978, virtually all trade in China was handled by a dozen foreign trade corporations (FTCs), under the direct control of MOFERT (Ministry of Foreign Economic Relations and Trade). This was a cumbersome system, since end-users rarely had a chance to meet and discuss requirements with suppliers. Restructuring of the system began after 1978 and a decade later was still in progress. The idea is to remove government and bureaucrats from commerce and industry, instead allowing managers and business people to take greater responsibility. They now have some control over any foreign currencies which they earn from sales abroad.

Enterprises can also purchase foreign currency through the banks rather than apply for allocations from the state plan. This is linked to the other reforms of industry, for instance enterprises being made increasingly responsible for their own profits and losses. Reforms in banking since 1979 have similarly attempted to increase efficiency, decentralisation, a wider range of financial services and increased savings.

Decentralisation of certain controls on trade after 1979 was thus part of the broader economic reform and was badly needed to help promote exports to pay for the mounting imports associated with modernisation although its progress has not been smooth. A growing number of provinces have been permitted to establish their own independent import–export corporations, a process begun in the early 1980s with Guangdong, Fujian and the three central municipalities (Beijing, Shanghai and Tianjin), and now including even far west regions such as Xinjiang. Whilst there have been various political factions attempting to reverse or slow down these types of decentralisation, the process seems to be progressing inexorably.

CHINA'S TRADING PARTNERS

The disagreeable experiences of heavy reliance on the Soviet Union and East Europe in the 1950s mean that China will probably never again depend on so few trading partners for the supply of important machinery and equipment. The share of communist countries (especially the Soviet Union) in China's trade in the late 1950s was almost two-thirds of the total. A more normal, multi-partner mixed trade is developing and is likely to be maintained, enhanced by diplomatic links. (The record of China's foreign relations and trade patterns in the past clearly reveals the interaction between trade and diplomacy.) The People's Republic of China's first trade partners were limited mainly to the communist countries. The break with the Soviet Union in 1960 and the increase in the number of countries recognising China in the 1970s, led to a marked increase in non-communist trading partners. This has accelerated during the 'open door' policy era.

Trading partners in the period 1960 to 1976

The reorientation of China to free its economy from Soviet influence after 1960 led Japan to replace the Soviet Union as China's main trading partner in 1965. During the period 1965–70, China basically returned to trade patterns which had existed in the mid-1930s. Japan, Hong Kong and Malaysia–Singapore became its main partners; others included Australia, Canada, France, West Germany and the United Kingdom. The United States was the exception; at that time it refused to trade with a communist country. The share of communist countries in China's trade fell to just over one-fifth; reintegration with the West was progressing.

In the period 1971–5 (the Fourth Five-Year Plan), China's trade rose rapidly, particularly with the United States with whom trade increased enormously after the removal of US government restrictions. By 1974, the USA was second only to Japan among China's trading partners. The share of communist countries rose again in the early 1970s but fell back by 1974, again to only about one-fifth of the total. Its most important trading partners in 1973 and 1974 were (in overall trade terms) Japan, the USA, Hong Kong, West Germany, Malaysia–Singapore, and Canada, followed by Australia, France, the United Kingdom, the Soviet Union and others.

Trading partners in the 1980s

In the post-Mao era, China has been trading widely and wishes to extend its links with the West, with communist countries and with the Third World. By 1983, the rank order was largely the same as in 1973, although Malaysia and Singapore's contribution had fallen to roughly that of the United Kingdom. A similar pattern is seen for 1985 (Table 9.1) and 1986, although in that year exports to the United Kingdom increased, as did imports from many other countries.

Table 9.1 China's major trading partners, 1985

Rank	Exports to	Export value (millions of US$)	Per cent of total exports
1	Hong Kong	5,746	22.17
2	Japan	5,609	21.64
3	USA	2,653	10.23
4	Singapore	2,093	8.08
5	Jordan	1,060	4.09
6	USSR	968	3.74
7	West Germany	678	2.62
8	United Kingdom	451	1.74
9	Brazil	422	1.63
10	Switzerland	352	1.36
	Total	20,032	77.30

Rank	Imports from	Import value (millions of US$)	Per cent of total imports
1	Japan	10,825	31,53
2	Hong Kong	5,148	14.99
3	USA	4,373	12.74
4	West Germany	2,394	6.97
5	United Kingdom	975	2.84
6	USSR	913	2.66
7	Canada	872	2.54
8	Australia	870	2.53
9	Italy	614	1.79
10	France	592	1.72
	Total	27,577	80.33

Source: Almanac of China's Foreign Economic Relations and Trade 1985.

It is useful to look at the balance of trade (i.e. the difference in the value of imports and exports) with different countries or groups of countries. In 1983, China basically seemed to be running a deficit with the developed world and a surplus with the developing world. It imported far more from the USA, Japan, Canada, France and Australia than it exported to them (trade with the United Kingdom showed a very slight surplus). By contrast, it exported to developing countries more than double the value of its imports from them. Hong Kong is an especially important export market for China (the biggest in 1986), but it re-exports many Chinese goods to the rest of the world. However, on balance, Japan was clearly China's most important trading partner. The European Community is of growing, if relatively small, significance.

In 1985, Japan was again the largest single exporter to China, followed by Hong Kong and the USA (Table 9.1). This ranking also mirrors the sources of loans and direct foreign investment (shown in Table 9.3). Japan (a major beneficiary of the imports boom in 1984–5), saw its exports to China grow by 64 per cent, but this trade had dropped by 20 per cent in 1986. Hostility to Japanese economic expansion, similar in many ways to that witnessed in Europe and the USA, may help other nations who wish to develop trade with China.

China has a special trade relation with Hong Kong and enjoys a large trade surplus with the territory. In 1984 the surplus was HK\$ 17.5 billion (US\$ 2.2 billion), enough at the time to cover a substantial part of its trade deficits with other countries. In 1986 China's exports to Hong Kong increased by 32 per cent over 1985, but much of this volume probably stemmed from goods originating in Hong Kong exported to China for processing and then re-exported to Hong Kong. The extent of such double counting is unknown but illustrates a problem in investigating China's trade with the territory.

The nature of the trading relationship between the two and the goods involved have changed over time. Hong Kong is growing in importance as a staging post for goods into and from China: for many years, about one-third of Hong Kong imports from China were for re-export, but the figure in the mid-1980s rose to half. Apart from providing trans-shipment services, Hong Kong's efficient banking system handles transactions for China and its indirect trading partners, with whom China may currently have no diplomatic ties. In 1987 almost 80 per cent of Hong Kong's entrepôt trade was related to China, either as a market or as a source of supply. The territory's trade with China grew in value by over 1,000 per cent between 1979 and 1988. Since 1985, China has been Hong Kong's major trading partner, with Hong Kong now as China's biggest export market.

BALANCE OF TRADE SINCE THE EARLY 1970s

The first oil price rise in 1973 affected China badly as it did many other countries both rich and poor, and its terms of trade deteriorated as the price of many imports rose considerably. Trade deficits (imports costing more than the value of exports) occurred in 1974 and 1975. Mao's death in 1976 led to the temporary suspension of many trade negotiations and these factors together slowed the growth of trade, the total of which remained virtually the same in the period 1974–9.

With the 'open door' policy after 1976, trade started to grow substantially. A massive expansion of imports occurred, with growth in real terms (not merely caused by inflation) of 32 per cent in 1977, 51 per cent in 1978 and 21 per cent in 1979, leading to balance-of-trade deficits in 1978–80. Deficits such as these contributed to the 'readjustment' of China's economic policy decided upon in 1978, involving a switch from heavy to light industry and agriculture, and an emphasis on balanced development, rather than rushed growth through imports. This last has hardly been achieved. In trade terms, large-scale direct imports of new technology were discouraged in favour of compensation (barter) trade and joint ventures, discussed later, which reduced expenditure of valuable foreign exchange. There was a push to earn via exports from the new light industrial sector.

However, since 1983, when a small trade surplus was recorded, growing and potentially serious trade deficits have recurred, reaching US$ 7.6 billion by 1985 according to some figures (US$ 13.7 billion according to others) and US$ 12 billion by the end of 1986. Whilst stated foreign trade policy is to balance imports against exports, in late 1984 the government stimulated a large growth in direct consumer goods imports. This was copied by local authorities nationwide, but especially in the south. The aim was to force some of the nation's huge accumulated savings into circulation and stimulate a rise in productivity and quality of Chinese-made goods, by exposing domestic producers to foreign competition and by giving workers the incentive of having consumer goods to buy. The resultant drastic imbalance of trade triggered firm government action in 1985 to halt the dwindling of foreign exchange reserves. It also caused much circumspection. Such an experiment, allowing imports to run out of control, is unlikely to be repeated. In the late 1980s it seems that the authorities were, if anything, trying to slow trade and restrict imports, and trade balances improved in this period.

Exports

Exports produce over 80 per cent of China's foreign currency earnings, with the remainder from tourism, labour services and remittances from Overseas Chinese. Exports have increased from US$ 552 million in 1950 to US$ 25,915 million in 1985 and around US$ 30,000 million in 1986. There has been not only a marked increase in the value of exports but also a structural change (Figure 9.2a). Agricultural products have declined from 55.7 per cent of the total value of exports in 1953 to 17.5 per cent in 1985. Light industrial products have increased from 26 per cent to 39.7 per cent, and heavy industrial products (including petroleum products) have increased from 17.4 per cent to 42.8 per cent. The major source of exports in 1985 was the coastal provinces (Figure 9.3). Most exports came from either the coastal provinces or central municipalities with a strong

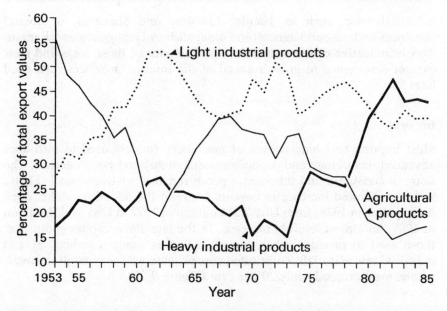

Figure 9.2a Commodity composition of exports, 1953–85

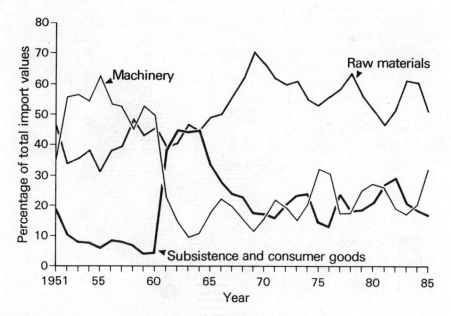

Figure 9.2b Commodity composition of imports, 1951–85

industrial base, such as Tianjin, Liaoning and Shanghai, or coastal provinces with a sound agricultural base, such as Guangdong and Jiangsu. This is indicative of the economic predominance of these areas and their current resurgence in growth ahead of the interior provinces, discussed later.

Imports

Most imports to China consist of machinery (mainly used to introduce advanced technology and production equipment) and raw materials. The share of consumer and subsistence goods remains relatively small (Figure 9.2b). A marked increase in imports followed the adoption of the 'open door' policy in 1978, from US$ 7,214 million in 1977 to US$ 34,331 million in 1985, an almost fivefold increase. In the late 1980s capital goods (i.e. those used to produce other things), including modern technology and industrial raw materials, constituted approximately 80 per cent of the total, with consumer goods only 20 per cent (Figure 9.2b).

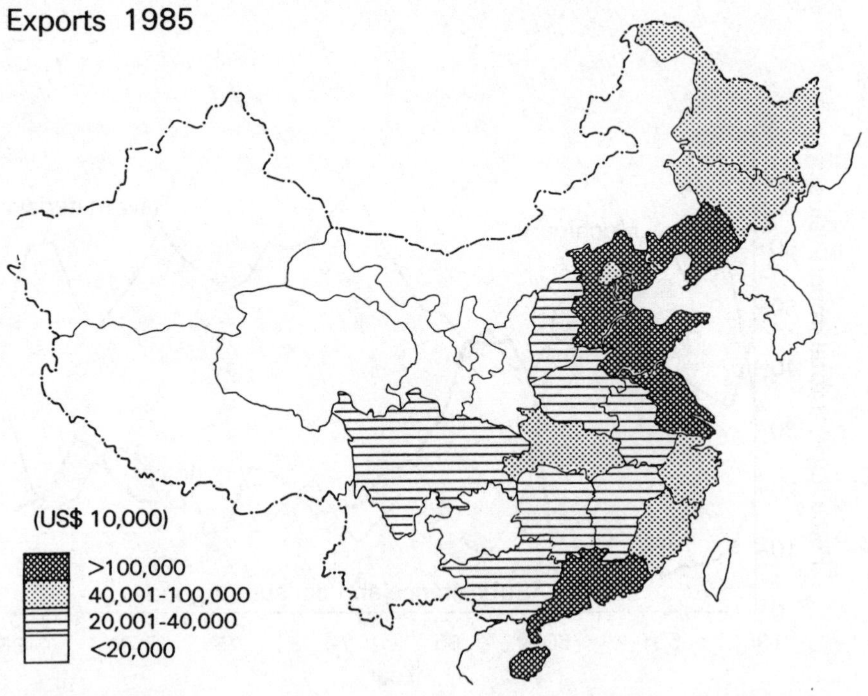

Exports 1985

(US$ 10,000)

>100,000
40,001-100,000
20,001-40,000
<20,000

Figure 9.3 Exports by province of origin

FOREIGN INVESTMENT

The use of foreign capital is a means of overcoming shortage of domestic funds or of readjusting the fiscal orientation of the economy. Foreign capital includes funds from other governments, commercial banks, enterprises and individuals, as well as international financial institutions and money markets (Table 9.2). In the case of China, it also includes funds from Hong Kong and Macao. Foreign capital mainly takes two forms: loans and direct foreign investment. It may also help to improve Chinese management, productivity and competitiveness so that more may be exported to earn foreign exchange and increase employment.

Borrowing and receiving capital

In the early post-liberation period, China borrowed from the Soviet Union and East European countries to build several hundred projects. In the early 1970s, chemical, metallurgical and energy production equipment was imported from capitalist countries. The increase in contact with the rest of the world since 1979 has included a number of forms of investment and co-operation in which foreign firms participate. The major forms of foreign involvement and trade with China are as follows:

Equity joint ventures (EJVs)

The foreign and Chinese parties to the enterprise pool money in agreed proportions. They invest in and operate the venture, sharing profits and losses in proportion to their equity stake. This is the form of investment preferred by China.

Wholly foreign-owned subsidiaries (WFSs)

Wholly owned subsidiaries of foreign companies and, since 1986, wholly owned companies, can be incorporated in China. The investor will be responsible for capital and input costs, marketing and management of the venture which runs for an agreed period. Profits after tax can be remitted overseas.

Contractual joint ventures (CJVs)

The Chinese enterprise provides the land, factory buildings and labour services for a foreign firm that provides equipment, capital and technical expertise. The period of investment is shorter than for the above methods, usually five to seven years, and the flexibility and short period make CJVs popular with foreign investors.

Joint developments

These represent a form of co-operation seen in the joint exploration for offshore oil. During the two stages of exploration and extraction the

Chinese and foreign partners take different levels of risk. For example, in the exploration stage, the financial and other risks lie with the foreign party, whereas in production, both sides contribute to the business.

Compensation trade (product buy-back)

The foreign investor receives payment in the form of the goods produced by the enterprise in which an investment is made. This is a form of counter-trade akin to barter, but it may also include payment in the form of other goods bearing no relation to the original venture or equipment supplied. It is popular with the Chinese government as it avoids excessive payments of foreign exchange.

Other forms of investment and co-operation

A number of other forms of investment exist. For example, leasing has developed since the early 1980s. Intermediate processing of goods has also grown, using raw materials, equipment and parts supplied to Chinese firms. (This is particularly important with regard to Hong Kong goods). There are various processing and assembly agreements under which foreign firms supply materials for Chinese enterprises to process or fabricate into finished products according to the foreign investors' designs and specifications.

This variety of forms of investment and co-operation indicates that China is seriously attempting to attract foreign investment and expertise. Since the adoption of the 'open door' policy, foreign capital worth some US$ 22,000

Table 9.2 Amount of utilised foreign capital 1979–85 (in millions of US$)

	Millions of US$	Per cent of type	Per cent of total
1 Loans from foreign governments	3,403.2	21.6	
2 Loans from international financial institutions	1,795.4	11.4	
3 Buyers' credit	760.4	4.8	
4 Loans from foreign commercial banks	8,209.2	52.2	
5 Others	1,561.6	9.9	
Total foreign loans	15,729.8	100.0	72.2
1 Equity joint ventures	1,007.8	16.6	
2 Contractual joint ventures	1,808.5	29.8	
3 Joint exploration	1,791.7	29.6	
4 Fully foreign-owned enterprises	111.0	1.8	
5 Compensation trade	869.0	14.3	
6 Others	473.0	7.8	
Total direct foreign investment	6,060.9	100.0	27.8
Total foreign capital	21,790.7	100.0	100.0

Source: Almanac of China's Foreign Economic Relations and Trade 1985.

million has been utilised (Table 9.2). 72.2 per cent of this has been in the form of external loans, whereas 27.8 per cent has been direct foreign investment. Loans from foreign commercial banks constitute the major share of foreign loans, with loans from foreign governments coming second.

Contractual joint ventures and joint exploration constitute almost 60 per cent of the total direct foreign investment. Foreign loans and direct foreign investment are generally utilised in different sectors of the economy. The majority of the external loans have been for mineral exploration, transport and heavy industrial development, whilst almost half of the direct foreign investment has been in offshore oil development and tourism.

Estimates of the relative numbers and value of foreign-funded enterprises established in China in the period 1979–85 suggest that there have been some 2,301 EJVs, with a total contractual value of US$ 3,377 million; 3,756 co-operative enterprises (value US$ 8,103 million) and about 120 wholly owned foreign companies, with a contract value of US$ 517 million.

The most popular form of investment by 1984 was in the form of CJVs, followed by EJVs, with joint exploration projects also of importance. The compensation trade form, whilst attractive in some ways, has not grown as fast, principally because of the difficulties of reaching agreement on product buy-back and the difficulties of selling some products on the international market where there is strong competition in price and quality.

There is a concentration of foreign loans and direct foreign investment from a relatively few countries (Table 9.3). Japan and the World Bank contributed almost three-quarters of total foreign loans in 1985, whereas Hong Kong and Macao were the sources of nearly half of the total direct

Table 9.3 Major foreign investors in China, 1985 (those with total investment of more than US$ 50 million)

Country (region) and international financial institutions	Total (millions of US$)	(%)	Foreign loans (millions of US$)	(%)	Direct foreign investment (millions of US$)	(%)
Japan	1,591.0	35.7	1,275.9	50.9	315.1	16.1
Hong Kong, Macao	1,016.4	22.8	60.7	2.4	955.7	48.9
World Bank	584.9	13.1	584.9	23.3	0.0	0.0
USA	381.6	8.6	24.5	1.0	357.2	18.3
West Germany	157.3	3.5	133.1	5.3	24.1	1.2
Britain	98.1	2.2	26.7	1.1	71.4	3.6
France	79.0	1.8	46.5	1.9	32.5	1.7
Total for these	3,908.3	87.6	2,152.3	85.9	1,7546.0	89.8
Total of all	4,462.1	100.0	2,506.0	100.0	1,956.2	100.0

Source: Almanac of China's Foreign Economic Relations and Trade 1985.

foreign investment, with the USA and Japan providing a further third. Outside funds are thus provided largely by Asian investors, financial institutions and governments. Investments from multi-national companies are relatively minor, although more visible in sectors such as offshore oil exploration.

LOCATION OF FOREIGN INVESTMENT

In addition to the legal and political reforms to ease the way for foreign investment in China since 1979, a number of geographical areas have also been designated as focal points in attracting such investment. While it is too simplistic to identify a few growth areas in a country that is witnessing such widespread developments, it is helpful to examine three main types of area: Special Economic Zones, open coastal cities and other areas wishing to attract foreign investment (Figure 9.4).

Special Economic Zones

Special Economic Zones (SEZs) were the first areas to be designated to attract foreign investment. Four were initially established in 1979: Shenzhen, Zhuhai, Xiamen and Shantou (Figure 9.4). (Hainan island was added as a fifth SEZ in April 1988, but discussion here concentrates on the first four). The SEZs represented a major initial attempt to attract foreign capital, investment, enterprise and technology, which would be located in 'controlled environments', strictly demarcated zones. They are similar in some ways to export-processing zones elsewhere in the Third World where an export-oriented industrialisation strategy is facilitated by providing legislative and tax concessions, infrastructure and serviced sites. China's SEZs offer a similar range of financial, legal and infrastructural induce-ments to foreign investors, including tax 'holidays', easy remittance of profits, and prepared sites with services and buildings. They also offer labour savings compared with many other countries in Southeast Asia plus, importantly, simplified bureaucratic procedures for investment, customs and immigration.

China's SEZs are a particular type of export-processing zone (EPZ). The SEZs had the initial objective of manufacturing to produce goods for export to earn foreign exchange but, very importantly, they have been regarded as social and economic laboratories, in which foreign technologi-cal and managerial skills might be observed and adopted, albeit selectively. They have had mixed success. For example, they 'export' more of their products into China than overseas: only 30 per cent of the SEZs' total products are estimated to be available for export rather than a desired 70 per cent. They have sometimes acquired dated, rather than the latest,

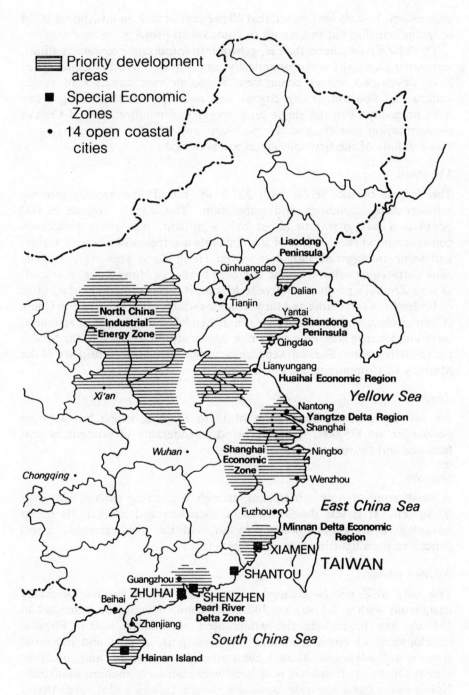

Priority development areas

■ Special Economic Zones

• 14 open coastal cities

Liaodong Peninsula

Qinhuangdao

Dalian

Tianjin

Yantai

Shandong Peninsula

Qingdao

North China Industrial Energy Zone

Lianyungang

Huaihai Economic Region

Xi'an

Yellow Sea

Nantong

Yangtze Delta Region

Shanghai

Wuhan

Ningbo

Shanghai Economic Zone

Chongqing

Wenzhou

East China Sea

Fuzhou

Minnan Delta Economic Region

XIAMEN

TAIWAN

SHANTOU

Guangzhou

ZHUHAI

Beihai

SHENZHEN

Pearl River Delta Zone

Zhanjiang

South China Sea

Hainan Island

Figure 9.4 Special Economic Zones, open cities and priority development areas

technology. It is also estimated that 40 per cent of foreign investment is not in manufacturing but in tourism or commercial projects.

The SEZs have, nevertheless, grown into important economic entities, somewhat physically and economically cut off from the rest of China. They have developed mixed economies, including vast amounts of retail, commercial and tourist capacity as well as simply manufacturing. They have probably been the single most researched manifestation of China's modernisation process and the problems and benefits accompanying it. Some details of the first four zones are provided:

Shenzhen

The largest of the SEZs, with 327.5 sq. km. It has rapidly growing infrastructure, buildings and population. The zone's original 30,000 population has grown to about half a million, plus many temporary construction workers. Most of the residents are from elsewhere in China, and some managerial staff come from Hong Kong and overseas. The zone's strategic position, on the northern border of Hong Kong, is crucial; as with Zhuhai's position relative to Macao, it forms something of a buffer or bridge between China and these colonies which will soon revert to China (Chapter eleven). Like the others, Shenzhen has a wide range of activities, including tourism, retailing and a new university. In the west of the SEZ is the industrial zone, Shekou, separately administered by a subsidiary of the Ministry of Communications in Beijing.

Zhuhai

An actively growing small zone of 15.16 sq. km, which has a target population of 175,000. It has received considerable investment in port facilities and tourist–recreational provision.

Shantou

A small zone of only 1.6 sq. km, though it is being extended to over 50 sq. km. It is less developed than Shenzhen and Zhuhai. Its major advantage is a good deepwater port but, with the areal extension, it will concentrate on light industry and agriculture.

Xiamen (Amoy)

The only SEZ not in Guangdong province, but in Fujian. Originally designated with a 2.5 sq. km industrial zone, it was soon extended to 131 sq. km to include the whole area of Xiamen island. Physical development of communications facilities, port, airport and industrial zones is well advanced. Xiamen has a prestigious university and numerous tourist functions. Emphasis is on relatively high-tech, modern industries. Xiamen also has a strategic location opposite Taiwan and it is felt that it may develop complete free-port, or similar, status.

The zones (particularly Shenzhen) have been criticised for not exporting as much as initially envisaged and for being avenues for the import of expensive foreign goods into China. However, their ambiguous and possibly over-ambitious initial objectives made it almost inevitable that there would be some disappointment and debate about their achievements.

Open coastal cities

During 1984 and 1985, in spite of some criticisms, the SEZs were being acclaimed by many as examples of what other Chinese cities could achieve. In Spring 1984, fourteen other coastal ports were opened for investment, the geographical spread of which would redress the southerly bias of the SEZs (Figure 9.4). These 'open cities' were to offer similar concessions to the SEZs, but while they are not provided with the same level of central-government funding for infrastructure development, 'Economic and Technical Development Zones' will be provided in them to attract foreign investment, often of a high-tech nature.

However, in June 1985 Deng Xiaoping started to voice misgivings about the SEZs and open cities, and it was announced that foreign investment was to be channelled into four of the biggest coastal cities – Shanghai, Tianjin, Dalian and Guangzhou – rather than all fourteen. They would grow in advance of the other ten, which would still be favoured but which, realistically, could not be pulled along at the speed of the four main cities. Whilst these were the better-known, arguably more attractive, cities, the growth potential of some of the others is also quite good and they may develop well in the future. In fact there are signs that, in at least some, the local administrations have continued to promote their cities' 'open-ness' in spite of central priorities.

It is interesting that many of the fourteen coastal cities and areas near the SEZs were previously ports in which foreigners had lived and traded with China. This comparison is something of a problem for the authorities, who are sensitive to suggestions of 'neo-colonialism' by Chinese critics and some foreign observers. Policy debates, description of the SEZs as being 'experimental' and the like have arisen over the past few years. Investment and corruption scandals involving Chinese officials and managers in the SEZs and elsewhere, such as in Hainan island, have also understandably made the authorities anxious about allowing too much free rein to those unaccustomed to it. Nevertheless, Shenzhen, Zhuhai and Xiamen are now well established and promise to grow quite strongly.

Other areas favoured for foreign investment

A number of other types of area are also keen to attract foreign investment, either with or without special inducements. In addition, in some areas being developed mainly for domestic purposes, investment and joint venture opportunities arise (see also Chapter ten):

The delta regions

Some broad areas are being favoured because of their accessibility and other reasons for foreign investment. Often, this involves development around an SEZ or a port city in recognition of the spin-offs and enlarged markets offered by such zones. In 1985 three 'open' economic regions were designated: the Chang Jiang Delta Economic Region (around Shanghai); the Zhujiang (Pearl River) Delta Economic Region (around Guangzhou), and the Minnan Delta Economic Region (around Xiamen). These are considered to have relatively advanced economic development and communication systems, and good tertiary education institutions. Such openings seem to be an attempt to spread benefits from the SEZs or ports to surrounding parts of the country.

Hainan island

This was, until 1987, in a category of its own, and offers certain inducements for foreign investment. It was promoted in conjunction with the fourteen cities in 1984, and in 1988 gained the status of a province, separating it from Guangdong. Its potential, amongst other activities, include tourism, agriculture and oil industry services. In April 1988 it was designated a fifth SEZ.

Cities granted economic autonomy

Some cities such as Guangzhou (Canton), Chongqing, Wuhan, Xian and Dalian have been given the right to cut a considerable amount of bureaucracy in getting their economic plans approved directly by the State Council. This means faster and surer approval of schemes and indicates greater commitment on the part of local officials to economic development and trade.

Certain peninsulas

It is reported that Shangdong and Liaodong (part of Lioaning province) peninsulas will be developed as integrated wholes, with enhanced potential for foreign involvement.

Domestic zones

A great deal of activity has been aimed at modernising certain areas of China for the domestic economy. Some such zones are based on natural resources, others on the existing volume of industry and economic activity. These include the North China Energy Zone, the Huaihuai Economic Region and the Shanghai Economic Zone which encompasses five provinces (Jiangsu, Zhejiang, Anhui, Fujian and Jiangxi) and the delta noted above (Figure 9.4). This zone provides more than a quarter of the state revenue and already handles over one-third of China's foreign trade.

This regionalisation of economic activity is aimed primarily at strengthening China's economic base rather than purely for attracting foreign investment (see also Chapter two). Nevertheless, in the upgrading, modernisation and general development of these regions and zones, there will no doubt be opportunities for foreign involvement in larger areas of China, with important long-term economic and structural implications.

SPATIAL IMPACTS OF FOREIGN TRADE AND INVESTMENT

Regional differences and spatial imbalances in development are now growing quite rapidly in China, though imbalances have been commonplace in the past. The urban and regional development policy of the Cultural Revolution contained a key component of 'self-reliance'. This involved the decentralisation and dispersal of economic activities so that cities and provinces, especially inland ones removed from the more dynamic eastern seaboard, could attempt economic self-sufficiency.

In terms of national strategy, this meant an almost total opposition to participation in a global network of foreign trade and investment, and the almost inevitable perpetuation of uneven geographical development of China that this would entail. The concept of self-reliance and equalisation of development has not entirely been dismissed, but it is now well recognised that China's search for modernisation will involve much greater collaboration with the rest of the world. Locationally-favoured sites are therefore expected to benefit disproportionately, as they did in the past.

Development on and away from the coast

Foreign involvement with China has a long and, at times, unhappy history. It includes the 'unequal treaties' of the nineteenth century which forced China to open up ports and trade avenues with the West (see Chapter three). In the years leading up to and including the Second World War Japan occupied parts of Manchuria as an industrial base and then invaded much of the east of the country. The major spatial effect of foreign involvement in China through the nineteenth and early twentieth centuries was to focus transport and development on to the eastern coastal provinces and the northeast. 'Decentralisation' and spread of development and economic activity from this eastern concentration was not a new concept of the Maoist era but it had also occurred during the First Five Year Plan (1953–7) and involved movement from larger to smaller centres and from coastal provinces to the interior (see Chapters two and seven). Indeed, of the twenty-two fastest growing municipalities up to 1958, all but one were in inland areas. The 'Great Leap Forward' (1958–60) continued the dispersion of economic activity and the rural industrialisation policy, though overall it had little effect in altering the relative shares of gross industrial output.

The policy of 'de-urbanisation' of the 1960s and 1970s has been reinforced by a more positive measure of fostering growth of small and medium-sized cities (see Chapter eight). There has still been a tendency for strict control of large cities but only since 1982–3 has the small-towns strategy been seriously challenged. This is because some fear that the suppression of agglomerative forces and 'natural' growth, and the promotion of economic activities in places lacking infrastructure and linkages, will hinder national development. Such a view acknowledges the importance of location, scale and accessibility in promoting linkages with the world economy. It seems to assure accelerated growth in places which have strong natural advantages (resources or accessibility) rather than in less well-endowed areas and settlements.

Policy for coastal growth

Although foreigners can now invest in most provinces, the designation of the SEZs, open coastal cities, and open economic regions has indicated the priority attached by the government to attracting foreign investment to these areas rather than elsewhere. The reasons seem obvious: agglomeration economies may develop; socio-political control of foreign involvement may be easier; and these areas have better infrastructure and links with the outside world, and ties with Hong Kong and Overseas Chinese in Southeast Asia. As mentioned, most of the foreign capital has so far come from Asia, and particularly from Japan and Hong Kong.

Investments are therefore very unevenly distributed spatially. Indeed, analyses of the distribution of joint ventures show that they are highly concentrated in the coastal provinces, particularly around large cities such as Beijing, Shanghai, and Guangzhou (Canton), and in the Special Economic Zones of Shenzhen and Xiamen (Figure 9.5). Although there are indications of a gradual diffusion of foreign investment into the interior provinces, the impact of foreign investment on most interior provinces to date has been small.

Capital from firms in Hong Kong and Macao is concentrated in the SEZs and nearby Guangdong and Fujian provinces. This is largely related to geographical proximity and ethnic ties which seem to play an important part in the locational decisions of foreign investors. Most of the investment from Southeast Asian investors (mainly by Overseas Chinese) is in Guangdong and Fujian provinces, and those involved seem particularly interested in investing in the districts from which they or their families originated. Their investments are particularly important in the SEZs and

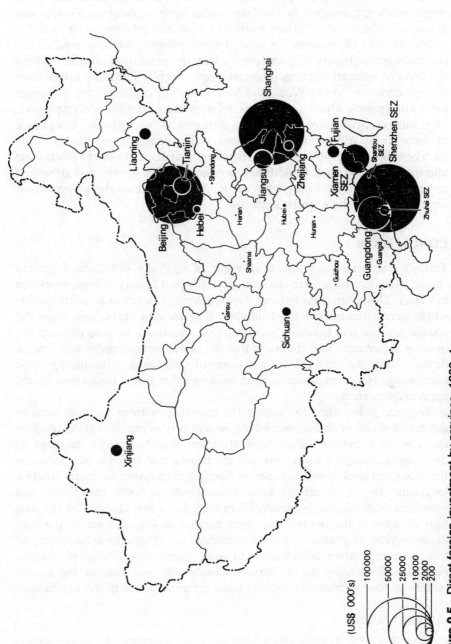

Figure 9.5 Direct foreign investment by province, 1982–4

small towns in Guangdong and Fujian, and especially the Pearl River Delta. By contrast, 50 per cent of North American and European investments are located in the large cities such as Beijing, Tianjin and Shanghai, where better urban facilities and industrial foundations exist.

The sources of exports are also therefore highly concentrated in the coastal regions (Figure 9.3) and particularly the main urban centres, where the level of industrialisation is already high. Such regions and towns have greater exposure to the West, and have more job opportunities in foreign joint enterprises which pay higher salaries than do Chinese employers. Migration controls have become less stringent in recent years, but people in the interior still cannot easily move to the coastal provinces to enjoy such benefits. This will tend to increase regional disparities between coastal and interior provinces. In addition, continued urbanisation and growth of large coastal cities will inevitably occur, fuelling the cumulative process of development of the eastern seaboard.

CONCLUSION

Today's tolerance and encouragement of spatially unbalanced growth should therefore be seen in the context of the past history of regional policy in China. This attempted to force development in locations and settlements which were often not the most suitable. It seems now that China hopes the coastal regions will become 'growth poles' which will help to develop the interior provinces. The theory is that economic development will 'trickle down' from the coastal to the interior provinces, stimulating their economies. However, there is little evidence that this is happening to any appreciable extent.

Regional policy aims to develop the interior provinces and link them to the modernisation and industrialisation which is so rapidly taking place in the coastal provinces. Particular attention must therefore be paid to developing transport links between the coastal and interior provinces, as the poor network presently deters foreign investment in many interior locations. Immense efforts have been made to shift manpower and resources to the west; however, there have been few changes in the past four decades in the overall east–west balance in China. Indeed, it seems likely in view of greater regional specialisation, competition and recognition of comparative advantage and scale economies, that regional disparities will increase in the future. The recent evolution of the spatial pattern of foreign involvement and trade certainly points to this conclusion.

FURTHER READING

Cannon, T. (1983) 'Foreign investments and trade: origins of the modernization policy' in A. Feuchtwang and A. Hussain (eds) *The Chinese Economic Reforms*, London: Croom Helm. A survey of the growth of foreign trade and investment in the post-Mao period, contrasted with policies for restricting it before the 'open door' policy.

Chu, D.K.Y. (1985) 'The trends and patterns of direct foreign investment in China', Centre for Contemporary Asian Studies, Working Paper No. 2, Chinese University of Hong Kong. Discusses the spatial distribution of foreign investment among other things.

Jao, Y.C. and Leung, C.K. (eds) (1986) *China's Special Economic Zones: Policies, Problems and Prospects*, Hong Kong: Oxford University Press. Collection of economic and geographic papers on modernisation and SEZs.

Kueh, Y.Y. and Howe, C. (1984) 'China's international trade: policy and organizational change and their place in the "Economic Readjustment"', *The China Quarterly* 100: 813–48. A detailed, technical paper on China's foreign economic relations, with special reference to 1970–83, the post-Mao era, incentives for foreign investment, and organisation of trade.

Phillips, D.R. (1984) *China's Modernisation: Prospects and Problems for the West*, Conflict Studies 158, London: Institute for the Study of Conflict. This paper reviews the process of China's modernisation and its implications for contact, trade and potential conflicts with the West.

Phillips, D.R. and Yeh, A.G.O. (1986) 'Special Economic Zones in China's modernization: changing policies and changing fortunes', *National Westminster Bank Quarterly Review*, February: 37–49. An accessible review of the SEZ policy and its implications for modernisation.

Phillips, D.R. and Yeh, A.G.O. (1989) 'Special Economic Zones' in D.S.G. Goodman (ed.) *China's Regional Development*, London: Routledge. Discusses the place of SEZs in modernisation and geographical balance.

Wong, K.Y. and Chu, D.K.Y. (eds) (1985) *Modernization in China: the Case of the Shenzhen Economic Zone*, Hong Kong: Oxford University Press. Discusses the nature of SEZs and their role in promoting modernisation, based on a detailed case study of the development of Shenzhen.

Case Study 9.1

THE OIL INDUSTRY

David R. Phillips and Anthony Gar-On Yeh

China may now be the world's largest producer of coal. It is also the sixth largest in oil output, with a significant share being exported. However, electrical power shortages remain one of the most serious obstacles to economic development and modernisation, acting as a brake on industrial growth. It is estimated that 25 per cent of industrial plant capacity is often idle because of power cuts. Why then does China strive to increase oil production, much of which is going for export?

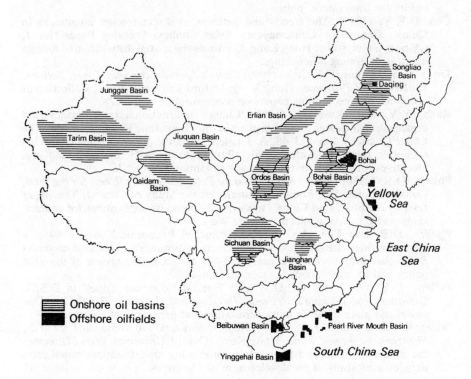

Figure 9.6 Oil resources, onshore and offshore

In the past, oil was a 'prestige' product and an important earner of foreign currency. The lowering of international oil prices in the mid-1980s led to almost a 50 per cent reduction in oil-export revenues, and China's oil-exporting corporation has been criticised for trying to sell in a stagnant international market when oil products and power have been in shortage at home. But there is still great activity in exploration for oil and development of oil reserves in China, in contrast to many other parts of the world.

China has a history as a substantial oil producer; oil was a major sector of Sino–Soviet collaboration in the 1950s and a reason for international technical and trade links even during the more isolationist 1960s. There are large onshore reserves, many of which are explored and exploited by fairly traditional methods. Some older onshore production areas, such as the Daqing field (in the northeast, producing half of China's oil in the early 1980s), were felt to have reached their peak output. However, in the offshore sector, advanced methods have been introduced, mainly with foreign assistance (see Figure 9.6).

Offshore oil

The annual output of China's oil industry reached 100 million tons in 1978, 130 million tons in 1985 and maybe 150 million tons by 1990. This increase is contrary to the predictions of many foreign commentators who had expected onshore production to dwindle. However, the ability of the petroleum industry to sustain long-term growth is doubtful, unless substantial offshore development comes about. At present, offshore production contributes only about 3 per cent of China's oil.

Nevertheless, in terms of foreign contact and trade with China, offshore oil has probably generated more excitement than any other single sector of the economy, but it does illustrate a number of the problems which China has faced in its modernisation drive. First, there has been the problem of attempting to link the 'open door' policy to a primary (mineral) product; second, the price of oil fluctuates considerably on the free market, rendering returns on investment uncertain; third, there have been considerable technical problems in the exploration and extraction in deep and inhospitable waters; fourth, question-marks exist about the ultimate volume of production; and, finally, China has had to rely extensively on foreign technology to be in a position to extract offshore oil.

It has for some time been recognised that China's continental shelf might contain much oil-bearing potential although the oil is difficult to locate and exploit. Offshore activity has been mainly associated with five areas: the South China Sea, South Yellow Sea, Beibu Wan, Yingge Hai Basin and the Bohai Sea (Figure 9.6). The China National Offshore Oil Corporation (CNOOC) was established in 1982 to monitor offshore exploration and development in what might have been a bonanza. China, in exploiting offshore reserves, has been anxious that a strong competitor such as Hong Kong or Singapore should not reap all the onshore benefits of the process. Production of equipment has therefore been located at Shenzhen and Hainan, for example, along with other onshore activities and services.

There is considerable commercial and political sensitivity and secrecy relating to offshore finds and potential, but China has been successful in attracting, if not always holding, considerable foreign participation in the offshore oil sector. The benefits of such participation tend to have been somewhat one-sided so far, but foreign companies may yet profit if substantial oil is found.

Lessons of the offshore oil exploitation

The Chinese offshore oil industry illustrates a number of features relating to modernisation and problems of the 'open door' policy. Many business people and commentators claim that China has sought excessive contract terms, 'over-exploiting' the eagerness of oil companies to invest for largely unknown returns. The negative foreign attitude to contract terms, operational conditions and impediments has almost certainly influenced attitudes to opportunities offshore and other forms of onshore investment.

The complex and often sluggish nature of Chinese bureaucracy is of particular significance to the oil industry because its capital-intensive nature renders it very vulnerable to bureaucratic delays and any financial penalties or duties on equipment. CNOOC holds a key role in negotiations with foreign firms but early on it was feeling its way, and the extent to which it could co-ordinate with other Chinese organisations was questionable.

The Chinese regard 'technology transfer', especially in the form of training, as an important aspect of the oil industry, presumably to be able to localise as much production as possible in the future. On-the-job training for Chinese technicians and labour sometimes meant an over-hurried replacement of expatriate workers and a loss of foreign control and influence locally.

Unattractive contract terms, strict policies of technology transfer, and high prices charged to the foreign firms for the purchase of goods and services have added to the 'up-front' costs which foreign firms have not been keen to incur as, if oil is not discovered, costs cannot be recouped. This situation has been made worse given that oil prices were falling, or were at best static, for much of the mid- to late-1980s, making offshore oil less attractive than previously. This illustrates the problem of relying on extraction of a mineral resource with fluctuating prices to finance modernisation.

Why, then, have foreign firms been keen to invest in China's offshore oil sector? The answer must be that large oil companies have been unwilling to be left out of one of the last major places in the world for new oil finds. This context makes their large investments seem more sensible: potential rewards are great and the dangers of being left out perhaps worse than the risks involved. The success or failure of the onshore and offshore oil industry in the late 1980s is crucial to the 'open door' policy and may well have a bearing on the whole process and speed of modernisation.

Chapter ten

The environmental impact of economic development: problems and policies

Bernhard Glaeser

The physical environment, in a broader sense 'nature', has been of basic importance for the economic development of all human societies. This has been true throughout the history of humankind; yet it is only recently that economic development began to threaten its own basis, i.e. the very existence of nature of which humans are a part. Does humanity have a right, perhaps a human right, over nature? This is one of the fundamental issues of environmental ethics, and under China's communist government it has sometimes been presented as society's destiny to have mastery over nature to fulfil human needs. Fortunately, there has been a growing recognition that this attitude leads to growth at the expense of the environment, producing a less secure basis for economic development and an unpleasant and unhealthy environment for the present and future.

PROBLEMS AND POLICIES 1949–77

Understandably, the initial concern of the communist government after 1949 was reconstruction and growth. The revolution, and the struggle against the Japanese invasion, devastated an already underdeveloped country peopled predominantly by impoverished peasants. The communists' position in the cities was rather weak, but it was urban dwellers who suffered most from the war, the destruction of industries and the ensuing economic crisis. The result in the cities was hunger, criminality and corruption. Under these circumstances, therefore, the Chinese government decided to rebuild industry and improve the social situation of the urban proletariat. Medical care was part of this scheme, and it was within this context that the concept of 'environmental hygiene' was developed. In particular, the provision of clean water was promoted as a means to combat the rapid spread of disease and epidemics in cities.

The environmental impact of industrial production was not recognised until the First Five-Year Plan (FYP) of 1953–7. Its main aim was to accelerate industrial production, particularly in heavy industry. This development pattern fitted the model of the Soviet centrally-planned economy and it was heavily financed by Soviet loans and aid. As a result, the urban population grew from 9.1 per cent of the total Chinese population in 1949 to 11.0 per cent in 1951 and to 12.7 per cent of the total Chinese population in 1957.

The First FYP did not mention the environmental impacts of industrialisation, though for economic reasons it advocated that natural resources should be used sparingly and that resources gleaned from industrial waste should be recycled. The principle of 'multi-purpose use' had an environmental impact and in the early 1970s it was introduced as a fundamental principle of Chinese environmental policy.

Research and laws under the First FYP

Scientific investigations in the 1950s showed that, even then, environmental problems were manifest. The main topics were air and water pollution, the ecological impacts of waste-water irrigation (which used effluent from industrial plants on the fields), environmental problems related to housing construction and city planning, and occupational diseases. In particular, the impacts of air pollution on health were studied in Shanghai, Beijing, Qingdao, Tianjin and the industrial cities of the northeast (Shenyang, Fushun, Anshan and Dalian).

According to some sources, the average sulphur dioxide concentration in the Tiexi industrial area of Shenyang was nine times higher than that in other areas. A paint factory in Dalian emitted chlorine gas whose concentration was 8.8 times higher than the threshold values (legal limits which should not be exceeded) established at that time in the Soviet Union. An investigation among pupils in Fushun showed that these children were ill ten times more often that children living in cleaner environments. Many 7–13-year-olds living in the vicinity of a steel mill in Shijingshan showed a tendency to develop enlarged livers. The illness rates in this area were attributed to high concentrations of sulphur dioxide.

In 1956, several regulations and new laws were enacted to protect the health of industrial workers and the urban population in general. This legislation ruled that industrial facilities should not be sited upstream or in the main-wind direction of residential areas. It called for the introduction of emission-abatement technologies and ruled that industries could not be sited in water-protection zones. These regulations were enforced by the local health authorities and some initial achievements were registered in the late 1950s.

From 'Great Leap Forward' to 'Cultural Revolution'

During 1958–60, the period of the 'Great Leap Forward' (GLF), environmental protection issues were somewhat neglected. Soviet development strategy was replaced by a home-grown Chinese variety which favoured combined agricultural and industrial development in rural areas within 'People's Communes'. But this campaign failed, in part because of the politically motivated withdrawal of Soviet aid.

Immediately following the GLF, more environmental regulations were drawn up, particularly some maximum-concentration threshold-value regulations. These were among the first such standards in the world. During the early phase of the Cultural Revolution (1966–9), the environmental concept established in the 1950s was basically abandoned. Development policy emphasised the benefits of self-reliance (see Chapter one), including much-reduced dependence on imports and stress on mass mobilisation. Environmental campaigns were nevertheless initiated; they also used mass mobilisations. One such campaign was called for by the Shanghai Revolutionary Committee to clean up the Huangpu and Suzhou rivers. This drive was more likely motivated by economic reasons, however, than by ecological ones:

> 90,000 persons were mobilized on the industrial and agricultural fronts in Shanghai to form muck-dredging and muck-transporting teams, waging a vehement people's war to dredge muck from Suzhow River. After 100 days of turbulent fighting, more than 403,600 tons of malodorous organic mire had been dug out.
>
> (*New China News Agency*: 21 October 1969)

In the early 1970s, still during the Cultural Revolution, there was a definite change in environmental attitudes and policy, probably influenced by discussions in the West such as the 'limits to growth' debate. Ecology became a political topic, and its emphasis shifted from issues of public health to ecosystem-related environmental protection. Accordingly, the term 'environmental hygiene' adopted from the Soviet Union was abandoned and replaced with the Western term 'environmental protection'. This development coincided with a more open political and economic outlook, which resulted in new imports of technology after 1972. In 1975, Zhou Enlai proclaimed the 'Four-Modernisations' programme in industry, agriculture, science and technology, and defence.

Campaigns on industrial waste and multi-purpose use

During the early 1970s, the campaign to recycle the 'three industrial wastes' (solid waste, waste water and waste gas) resurrected the concept of multi-purpose use from the 1950s. But its emphasis changed from an economic to an ecologically oriented principle. Multi-purpose use was enforced mainly in the steel, chemical, sugar and paper industries, and the media reported extensively on new recycling methods.

Some authors tended to interpret the successful recycling programmes and projects as part of the Chinese cultural heritage, as the traditional Daoist (Taoist) form of ecologically sound interaction between society and nature. According to this interpretation, recycling, for instance, as a basic principle of traditional agricultural technology (e.g. using human excrement as dung fertiliser), was transferred to modern industrial production.

In other sources, however, multi-purpose use was attributed less to Daoist philosophy and its emphasis on harmony with nature than it was to Marxist natural dialectics: environmental protection vanquished the contradiction between economy and ecology. Good and bad were considered relative values and the campaign of multi-purpose use propagated the tenets of 'converting wastes into valuables' and 'converting the harmful into the beneficial'.

It is obvious that Maoist ecological theory and campaigns were, to some extent at least, designed to serve the political and economic interests of the Maoist faction – that is, to strengthen an independent, autarkic economy. While lack of economic resources motivated the multi-purpose use campaign, it is doubtful, according to some experts, whether it contributed to abating the growing ecological crisis.

International participation and its consequences

In 1972, the first UN Conference on the Human Environment took place in Stockholm. Along with many other countries, China considered this conference to be the spring-board for its own national environmental policy. The Chinese delegation linked environmental problems to the problems of development:

> In the developing countries, most of the environmental problems are caused by underdevelopment. ... Therefore the developing countries must mainly direct their efforts to develop their national economy, build their modern industry, and their agriculture ... to adequately solve their own environmental problems.
>
> (Quoted in Sternfeld 1984: 22)

This formulation was changed slightly and added to Clause 4 of the final version of the UN Conference's Declaration on the Human Environment.

Whereas, in the past, environmental policy measures were motivated by economic *or* developmental needs, here environment *and* development were explicitly related to one another, and the solutions to problems in one area were inextricably linked with solutions to problems in the other. Historically, more development has usually meant worse environment, and examples of this can be found in China. On the other hand, the majority of developing nations – in Stockholm led by Brazil – argued that environmental damage was primarily caused by the industrialised countries who were consequently responsible for pollution abatement. Underdeveloped countries should not worry about environmental protection before they had reached the industrial wealth and living standard of the developed nations.

The link between environment and development must therefore be considered to be one of the origins of ecodevelopment – that is, ecologically based sustainable development – which was to be promoted by the UN Environment Programme (UNEP), the UN agency founded in Nairobi as a result of the Stockholm conference.

On a national scale, China began to implement ecodevelopment principles by convening the first National Environment Conference in 1973 in Beijing. Policy guidelines for environmental protection were drawn up at this conference; environmental administrations on the local, provincial and national levels were set up and environmental research institutions were established.

Qu Geping, then the Deputy Director of the Bureau of Environmental Protection of the State Council (who might be called the 'father' of Chinese environmental policy), stressed the preventive element to environmental protection in guidelines and policy statements. He maintained that environment and economic development had to be co-ordinated, which in fact is a restatement of the basic principle of ecodevelopment.

Types of environmental problems

Within the past thirty years, at least a quarter of China's forests have been felled. Erosion damage affects 1.5 million sq. km (i.e. one-sixth of China's total land area). Deserts have been encroaching steadily since 1949: from 110 million ha to 130 million ha. Intensive agricultural production techniques include the application of fertilisers and pesticides, which thus deposit harmful chemicals in soil and water. Finally, in the cities, air, water and noise pollution have reached discomforting levels. Only 8 per cent of all waste water is treated (Figure 10.1).

Air pollution

Air pollution, mainly in the form of smog and dust, plagues most urban areas and industrial centres (Figure 10.2). The major sources of air pollution are industrial processes and coal-based energy production. Another source, however, are the winds which carry dust from the deserts of Inner Mongolia and the loess plateau to the northern settlements (see Figure 4.7 on p. 95). Since the main cause of dust loss from the loess plateau is soil erosion resulting from deforestation, this type of pollution can be classified as of human causation as well.

In 1986, 24 million tonnes of sulphur dioxide and 22 million tonnes of dust (2.3 tonnes per sq. km) were registered; threshold values are

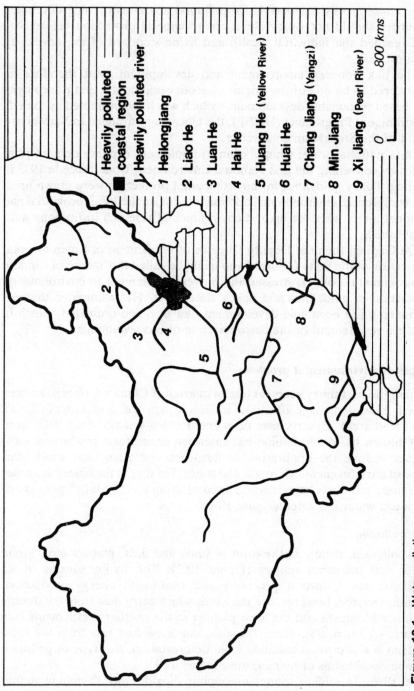

Figure 10.1 Water pollution
Source: After Schaffer (1986)

Heavily polluted coastal region
Heavily polluted river

1 Heilongjiang
2 Liao He
3 Luan He
4 Hai He
5 Huang He (Yellow River)
6 Huai He
7 Chang Jiang (Yangzi)
8 Min Jiang
9 Xi Jiang (Pearl River)

0   800 kms

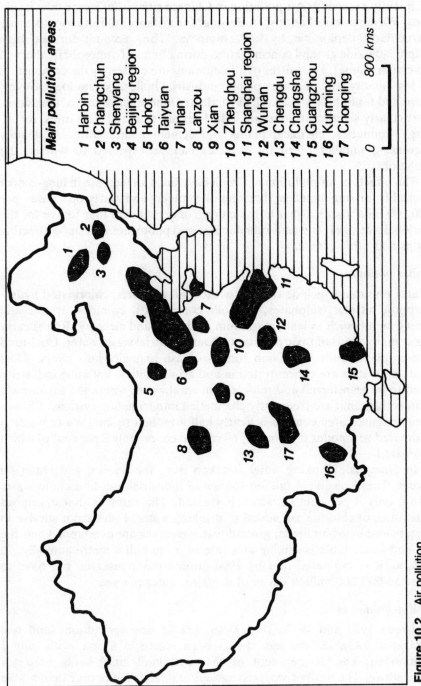

Main pollution areas

1 Harbin
2 Changchun
3 Shenyang
4 Beijing region
5 Hohot
6 Taiyuan
7 Jinan
8 Lanzou
9 Xian
10 Zhenghou
11 Shanghai region
12 Wuhan
13 Chengdu
14 Changsha
15 Guangzhou
16 Kunming
17 Chonqing

0   800 kms

Figure 10.2   Air pollution
Source: After Schaffer (1986)

frequently exceeded. Air pollution in Chinese cities is far worse than it is in most Western conurbations. In Beijing, for instance, the situation is made particularly deplorable by low chimneys. They account for the high sulphur-dioxide ground concentration during times of atmospheric thermal inversion (smog). Four-fifths of all emissions are caused by the combustion of low-quality coal in power plants, industrial boilers, steam locomotives, ships and household stoves. Heavy industry is another source of emissions. Particularly significant polluting industries are iron-and-steel mills, metallurgy, refineries, chemical plants and cement manufacture. A steadily increasing number of motor vehicles can be expected to worsen the situation.

The effects of air pollution in China have been an increase in lung-cancer mortality among humans (between seventeen and thirty-one cases per 100,000 inhabitants a year in the cities, compared with four to five for the national average), and acid-rain damage and photochemical smog affecting crops and other plants.

Water pollution

Water pollutants include organic wastes, oil products, chlorinated hydrocarbons, nitrates, sulphates, phenolic compounds, cyanides, arsenic and heavy metals such as lead, chromium, cadmium and mercury. Rivers carry these many kinds of toxic pollutants from industrial waste water. Coal-fired power plants emit 15 million tonnes of ash annually into rivers. Most polluted waters are concentrated in and around cities. Polluting industries tend to be transferred and relocated in rural areas, where the toxic waste waters they emit are frequently channelled into irrigation systems. China's total irrigated area comprises nearly half a million sq km. Waste water is estimated to amount to a total of 60 million cu. m, only 8 per cent of which is treated.

In Shanghai, drinking water is taken from the Suzhou and Huangpu rivers. These receive 4 million tonnes of industrial and household waste daily, only 4 per cent of which is treated. The result is that enormous quantities of chlorine are added to drinking water. Fish cannot survive in waters within urban limits, ground-water reserves are overtapped and the ground-water table is sinking at a rate of up to half a metre annually. In Beijing it is estimated that by 1990 ground-water reserves will have to provide 600–800 million cu. m of drinking water per year.

Soil degradation

Between 1957 and 1977, 210,000 sq. km of new agricultural land was reclaimed. Arable land has always been scarce in China, with only 1 million sq. km (10 per cent of the total land area) being used for agriculture. The newly recovered agricultural land is of poorer than arable quality, with yields only one-third of the national average.

In the same period, 330,000 sq. km (one-third of the previously cultivated prime arable soil) was lost, in part as a result of natural hazards but also as a result of intensified development in rural and urban areas, including the construction of irrigation canals, reservoirs, new roads, settlements and new factories. Population increased dramatically during this period.

The decrease in agricultural land has been aggravated by a decline in quality. 40 per cent of arable land consists of low-grade soils; 20 per cent is thin-layered hillside soil subject to erosion; 9 per cent is sandy soil; 8 per cent is waterlogged, lowland saline–alkaline soil; 3 per cent is low-grade paddy field. Such poor quality land has a yield capacity of only some 0.75 to 1.5 tonnes of grain per hectare per year, only about a quarter to half of the national average.

Some cropping practices have also contributed to soil deterioration. Careless irrigation, for instance, has increased the salinity and alkalinity of soils on the Hubei plain. It has also turned arable land in Jiangnan, Hunan, Jiangsu and Guangdong into bog or gley soils. Other adverse cropping practices include the failure to rotate crops (e.g. continuous, double- or triple-cropping of rice), replacing organic manure with chemical fertiliser, the failure to alternate between wet and dry crops, and the reduced application of green manure and nitrogen-fixing legumes. Deficient soils with a shallow layer of humus do not respond to increased inputs of fertiliser and water.

PROBLEMS IN THE POST-MAO PERIOD: INDUSTRY AND ENERGY

After 1978 the Chinese mass media began to publicise the environmental policy concept drawn up in 1973 at the First National Environment Conference. Reports of severe environmental damage began appearing in the media for the first time, as did other 'sensitive' issues, such as regional famines, unemployment and criminality.

Economic reform policies and the environment

The era following Deng Xiaoping's (second) rehabilitation after 1978 was characterised by the opening up of information and a more pragmatic, non-Maoist approach to development. In short, 'readjustment' became the catchword, heralding a new era of social and economic reforms. Environmental policy became an integral part of the Four Modernisations (of industry, agriculture, science and technology, and defence). An environmental protection clause was added to the Constitution, making protection of the environment a governmental obligation. Then in 1979 the Fifth National People's Congress adopted the Environmental Protection Law of the People's Republic of China for trial implementation. This law was:

established in accordance with Article 11 of the Constitution ... which provides that 'the State protects the environment and natural resources and prevents and eliminates pollution and other hazards to the public'.

(Environmental Protection Law, Article 1)

The function of the law is:

to ensure, during the construction of a modernised socialist state, rational use of natural environment, prevention and elimination of environmental pollution and damage to ecosystems, in order to create a clean and favourable living and working environment, protect the health of the people and promote economic development.

(Article 2)

Article 6 of the Environmental Protection Law incorporated a principle of Western environmental policy, namely, 'whoever causes pollution shall be responsible for its elimination'. Article 18 imposes a fine for exceeding pollution standards. Article 32, paragraph 1, imposes fines for violation of the law; paragraph 2 provides for making violation of the law a criminal offence.

While all of this sounded very promising, a major problem arose familiar to Western observers: its implementation. This difficulty was publicised in the press. Consequently, in 1981 the adjustment policy was strengthened and environmental protection was made a mandatory part of development policy. The areas in which the law was most strictly enforced suggest where the leadership thought the main sources of environmental problems existed. Between 1980 and 1983, environmental impact statements were written for sixty construction projects in heavy industry and manufacturing, mining, water management and water-related construction projects including canal digging and power-plant construction, city-planning projects and urban renewal. Urban development and urban-renewal projects began to include the installation of waste-treatment facilities. Fines were levied for emitting pollutants above a certain standard and for improper disposal of solid and liquid wastes. Fines were reported not only against production units but against managers as well, a further feature of the economic reform principle of direct responsibility.

Problems and policies

Solving environmental problems cannot be left to individual or autonomous enterprises; it must be part of a national movement which includes a broad spectrum of reforms, not limited to government, ranging from the political sphere to education. The industrial reforms undertaken since 1980 have tended to increase enterprise autonomy, producing contradictory effects on controls of environmental pollution. Enterprise managers can be held responsible; yet their search for profits (and competition with rivals) reduces the desire to spend on controlling emissions.

The results of inefficient implementation of environmental policy can be seen clearly in China's urban industrial areas. Here, air pollution, the most prominent environmental problem, is related to the reliance on coal for energy. Although China's annual average energy consumption amounts to some 850 million tonnes hard coal equivalent (HCE) (not even double West Germany's annual average energy consumption) dust emissions in China, amounting to 14 million tonnes annually, are more than fifteen times higher than West Germany's. China's sulphur-dioxide emissions, at 15 million tonnes per annum, are four times higher than West Germany's. This much worse emission is in part due to the quality of coal used, where coal combustion represents 51 per cent of primary-energy consumption in that country. On the other hand, overall energy-use efficiency in China is very low, only about 27 per cent compared to 40 per cent or more in Western industrialised countries. For the Chinese leadership, this discrepancy demonstrates the clear need for industrial modernisation. Moreover, it explains why economic inefficiency in the industrial sector is so strongly

Table 10.1 Data on the air quality of Chinese cities

Location	SO_2 (mg/m³)	Dustfall (t/km²/ month)	Dust concentration (mg/m³)	Type of data
Beijing	0.20	38	0.87	Average for the heating period in 1978.
Shanghai	0.08–0.17	44–1,822	0.28–0.54	Yearly average for 1977 in different parts of town.
Shenyang	0.087	38.5	0.512	Yearly average for 1978.
Chongqing	0.21–0.78	13.5–30.5	0.46–2.02	Daily averages for February and March 1980.
Lanzhou	0.25	35.9	1.32	Average for the heating period in 1978.
Taiyuan	0.24	35	0.2–19.6	Yearly average for 1976 (daily averages for 1977–9).
Nanjing	0.094	63	0.11–0.28	Daily averages for 1976.
Xuzhou	0.150	47	1.26	Average for the heating periods from 1975 to 1978.
Fushun	0.05–0.11	48–63	0.96–2.5	Daily averages for 1974.
Standards				
China	0.150	6–8	0.150	Long-term exposure (daily averages).
West Germany	0.140	10.5	0.100	Long-term exposure.

Source: Kinzelbach 1987: 179, in Glaeser, B. (ed.) 1987.

linked to implementation problems in pollution abatement efforts: in China, energy is wasted.

The consequence of low energy-use efficiency is heavy urban air pollution (Table 10.1). The main air pollutant is dust: dustfall and dust quantities exceed environmental standards just about everywhere. Soot and dust are the main sources of polycyclic aromatic hydrocarbons, particularly benzo-α-pyrene, a highly carcinogenic compound. In Beijing during the Winter heating period the benzo-α-pyrene content of the air reaches 6.9μg per 100 cu. m. The next worse air pollutant is sulphur dioxide, the concentration of which exceeds emission standards in a number of cities.

Air pollution poses a severe health hazard in China, causing such ailments as lung cancer, cardio-vascular disease, chronic bronchitis and chronic nose and throat infections mainly in northern Chinese cities, and silicosis in mining areas. Any comparison between China and Western countries regarding the air quality in urban centres is bound to prove unfavourable to China.

PROBLEMS AND POLICIES FOR THE RURAL SECTOR

For nearly 4,000 years, Chinese farmers have practised highly sophisti-cated, intensive agriculture, including the efficient use of water and irrigation, plant and animal recycling systems, composting, biological and mechanical pest control, waste management and faeces recycling. Despite access to traditional and modern agricultural knowledge, serious ecological damage persists in China. Forests decline despite afforestation pro-grammes; fertile land continues to decline as a result of poor water management, salinisation and overfertilising. There has been a rapid increase in the use of chemical fertilisers and pesticides.

Present efforts in agricultural development and environmental protec-tion include land improvement and better drainage to avoid salinisation. it seems that these problems have been aggravated by the contract farming system, in which the family functions as the basic production unit. It should be asked, therefore, what environmental policies have contributed to solving these problems, particularly in forest management, pest manage-ment, and bio-gas production?

Forest management

Only about 12 per cent of the total land area in China is covered by forest. As a consequence, firewood, the most important fuel in rural areas, is scarce. The forested area of Yunnan, a major reserve, declined by 13 per cent over a period of ten years. Forests in Sichuan have declined by as much as 30 per cent since the 1950s. Deforestation in mountainous regions has led to increasingly severe soil erosion. The Chang Jiang (Yangzi) River is turning murky-yellow from the mud of washed-away topsoil. The Yellow River has been transporting huge quantities of topsoil eroded by water for centuries. This loss of topsoil implies a general deterioration in agriculture because it cannot be compensated for by increased applications of fertilisers.

Afforestation and reforestation programmes have been pet policies of successive Chinese leaderships since 1949. The plans are to increase forested areas by 20 per cent (66 million ha of new forest) by the year 2000. The regions of special attention are northeastern China, Inner Mongolia, plus Sichuan and Yunnan in the southwest. Combined, these areas contain 50 per cent of the total forested area in China and 60 per cent of the timber reserves. The afforestation programmes are designed to fit into the overall agricultural land-use planning scheme: 40 per cent of the forests should be in mountainous regions, 20 per cent in hilly areas and 10 per cent in the fertile plains.

It is doubtful, however, whether such ambitious plans will come to fruition. The original 1978 plans had to be modified and by 1979–80 had been significantly cut back. The growing rural population required more arable land and firewood than was compatible with the original afforestation plans. It has since been reported that even local authorities no longer observe all environmental protection regulations.

Pest management

In the late 1970s Western agricultural delegations to the People's Republic discovered that agricultural practice in China included an advanced form of pest control – namely, integrated pest management (IPM). IPM uses a combination of various biological, chemical and physical techniques in conjunction with traditional farming practices. In a broader sense, plant breeding, cultivation techniques and crop rotation are integral parts of the IPM system.

Between 1949 and 1958 traditional agriculture was practised in China with virtually no developed programme of pest control. DDT was in use, but only to control insects directly harmful to humans. Between 1958 and 1970 chemical pesticides dominated the agricultural pest-control scene; they were in widespread use, particularly on collective farms. New varieties of rice had been introduced in China after 1970 to boost rice production. However the widespread use of chemical pesticides had eliminated the natural enemies of the plant hopper (an extremely destructive pest) during the 1960s, and rice production fell into a crisis between 1970 and 1972. Similar problems occurred with cotton-growing and, as a result, economic and ecological arguments combined in a truly ecodevelopmental effort led to the search for an alternative approach to pest control.

In 1979, IPM in China was applied over the largest area of any country in the world. Biological-control techniques such as the breeding of *Trichogramma*, a parasitic wasp, or *Bacillus thuringiensis*, a pest-killing bacterium, were leaders on the international scene. But the concept of agricultural modernisation with its emphasis on chemicals, electrification, irrigation and mechanisation had not been abandoned. Although IPM and modernisation coexist, the return to family contract farming is gradually eroding the institutional basis for IPM because it can only be applied effectively over large areas.

Bio-gas and energy supply

'Bio-gas' is largely methane, a product of the decomposition of organic matter. It can be produced in a controlled and useful manner in specially constructed concrete containers which have been sunk into the ground. The flammable gas given off when crop waste, animal and human excrement put in it break down under proper conditions may be used for cooking and light. The residual organic sludge is valuable manure. Bio-gas plays only a minute role as a source of energy in China: estimates suggest 1.6 million tonnes hard coal equivalent per annum: only 1.9 per cent of total energy supply. The authorities, however, including the Ministry of Agriculture, rate bio-gas very highly as a source of rural energy and a valuable substitute for straw, coal and firewood. It can, therefore, contribute to the protection of forests and other environmental goals.

In 1978 the existence of 7 million bio-gas digesters was reported. Of these, 5 million were located in Sichuan. A strategy to popularise bio-gas followed, with the aim of getting 70 per cent of households to operate digesters and establishing 'excrement management teams' who were responsible for building, repairing and emptying digesters. The bio-gas campaign was easy to carry through in the south of China, but in the north, where dry agriculture (using dry manure) is practised, it met with passive resistence. However, the future of bio-gas is uncertain in spite of its achievements, given the new agricultural policies which have individualised farming. According to some reports, funds are drying up even for successful waste-management teams. New high-quality digesters requiring little or no maintenance, which can be operated by rural households without technical assistance, are apparently to be produced industrially. The more extensive use of bio-gas would have important implications for economic, energy and environmental policy.

CONCLUSION

China has suffered severe environmental problems in urban industrial and rural areas, and has pursued an environmental policy which has remained surprisingly consistent despite significant changes in the political situation. Unfortunately, it was also consistent in its inability to implement fully all of the environmental measures.

The development approach must be considered at least to a degree as an ecodevelopment approach (i.e. involving environmental sustainability), whether intentionally or accidentally, because it held to the conviction that environmental considerations were an integral part of development policy and, ultimately, economic and ecological issues could not be reasonably separated. This conviction was obvious in important aspects of rural development, illustrated by the integrated pest-management programme, energy policy and bio-gas development. It was less obvious in urban industrial policy, during the earlier stages of which environmentally sound strategies such as waste recycling and energy saving were strongly promoted for economic reasons. As public health deteriorated and working conditions steadily worsened, and as environmental hazards and the effects of environmental damage in rural and urban settings made themselves felt, it became clear to the leadership that neglect of these factors also had serious implications.

Consistency in environmental policy despite political changes can be traced in part to the Chinese heritage and traditional way of thinking. It was, in fact, not so very long ago that China was an agrarian society completely dependent on natural recycling, the re-use of agricultural wastes, the use of organic manure fertilisers, biological pest control, multicropping and low-energy inputs. This approach to agriculture in the sense of re-use or multi-purpose use of waste, served to some extent as a guideline for industrial production.

On the other hand, environmental policy has never been fully implemented, despite intentions, consistency and bureaucratic institutionalisation. The country simply lacks the financial resources to install expensive technology on a large scale. There is little incentive at factory level to implement costly policies. There is insufficient enforcement of pollution fines, because this would involve the state, as property owner, taking money out of one pocket and putting it in another. Bureaucracies are not usually made to implement novel strategies. Successive failures of economic reforms suggest that not much more can be expected of environmental policy. A well-planned and sophisticated environmental protection policy is useless if, at the same time, economic reform abolishes the very institutions needed to implement policy measures. For instance, IPM and, to some extent, the bio-gas programme were dependent upon the rural collectives which were abolished in the rural reform.

Such counter-productive effects will have to be eliminated if Chinese environmental policy in its ecodevelopment approach is to be implemented successfully. Aware of this, the Director of the Environmental Protection Bureau, Qu Geping, argues for stronger environmental management at all levels of administration. This should include an environmental protection responsibility system designed to motivate provincial governors, mayors and other leaders down to factory heads to take action. He also proposes a centralised policy for the control of pollution of the air and water and of the disposal of solid waste.

ACKNOWLEDGEMENT

I am grateful for the invaluable assistance of Mary E. Kelley-Bibra in ironing out some of the rough spots in this chapter.

FURTHER READING

Boxer, B. 'China's Three Gorges dam: questions and prospects', *The China Quarterly* (1988) 113: 94–108.

Glaeser, B. (ed.) (1987) *Learning from China? Development and Environment in Third World Countries*, London: Allen & Unwin.
   This features Chinese environmental policy and planning in agriculture, industry, energy production and human settlements. Environmental policies are related to the corresponding strategies for development. Links are drawn to other developing countries which may, to an extent, use China's experiences.

King, F.H. (1949) *Farmers of Forty Centuries, or Permanent Agriculture in China, Korea and Japan* (fourth impression, edited by Bruce), London: Jonathan Cape.
   First published posthumously in 1911, this is a classic in agricultural literature. Mrs King travelled through Asia in 1909, describing ecological techniques and farming methods, such as the recycling of agricultural wastes, conservation of soil fertility and organic manuring as they have developed over 4,000 years of agrarian life in East Asia.

Lieberthal, K. and Oksenberg, M. (1988) *Policy-Making in China: Leaders, Structures and Processes*, Lawrenceville: Princeton University Press.
   Examines the evolution of post-Mao policy towards three major energy projects: Shanxi coal, South China Sea petroleum and the Three Gorges dam on Yangzi.

Pannell, C.W. and Salter C.L. (1985) *China Geographer 12: Environment*, Boulder, Colorado: Westview Press.
   A collection of papers by Chinese and American geographers, including problems of the loess region, agriculture, urban environmental quality, Beijing's water problem, fisheries, wildlife protection and others.

Qu Geping and Lee Woyen (eds) (1984) *Managing the Environment in China*, Dublin: Tycooly Press.
   A reader on the environment by Chinese authors. It gives a broad overview of environmental problems and the protection measures enacted and implemented to tackle them.

Ross, L. and Silk, M.A. (1987) *Environmental Law and Policy in the People's Republic of China*, New York: Quorum.
   Not yet seen at time of going to press.

Smil, V. (1984) *The Bad Earth: Environmental Degradation in China*, Armonk, New York: M.E. Sharpe Inc. and London: Zed Press.
   After his well-known book, *China's Energy* (1976: Praeger), Smil wrote a report on the environment in China for the World Bank which became the

core of *The Bad Earth*. It gives a critical and pessimistic evaluation of China's situation in regard to land, water, air and related environmental problems, including the extinction of species, and the conflicts between city and country-side. (The title is a play on that of a famous 1930s' novel by Pearl Buck – also made into a film – based on peasant life in China, *The Good Earth*.)

Chapter eleven

Defining and defending the Chinese state: geopolitical perspectives

Alan Jenkins

Do aspects of the geography of China make defending its territory particularly difficult? Have China's leaders since 1949 used force to try to claim particular territories, and if so what areas have they prized? What has been the country's role as a global and regional power? Has its foreign policy changed significantly since 1976? These are the questions considered in this chapter.

The approach will be essentially geopolitical. Geopolitics is a branch of geography or military analysis which considers how geography shapes leaders' policies with respect to the state's internal and external security. For example, Beijing's policy towards Tibet can in part be explained by the problems posed by distance, mountain and desert barriers and ethnic differences.

A geopolitical perspective can help us to understand aspects of contemporary China. However, care must be taken in using this approach. At worst we can fall into a geographically deterministic perspective, assuming that 'geographical factors' are the only determinants of state policy, while neglecting the role of culture, history, ideology, social structures, economics and so on. For example, in the case of Tibet, 'geography' does not explain why Chinese leaders have long seen Tibet as part of China and been willing to use force to assert control. Thus we have to recognise the complexity of the issues involved: the geopolitical lens through which we view them is powerful but partial and perhaps distorting.

Wider questions of Chinese foreign policy will be only briefly referred to, though in contrast to other chapters which emphasise events after 1976, this chapter will consider the whole period since 1949. (For a summary of developments since 1949 see Table 11.1). This is because it can be argued that for internal security and foreign policy, there are fundamental continuities in the whole period. Indeed, a geopolitical perspective would suggest that the durability of geographic factors helps explain that continuity.

Table 11.1 Security issues, 1949–89

1949	The Communist Party forces of the People's Liberation Army (PLA) defeat the forces of Chiang Kai-shek and his Guomindang (Nationalist) Party. The People's Republic of China (PRC) is proclaimed on 1 October. Many Guomindang supporters escape to Taiwan and proclaim their own government of China as 'Republic of China', which gains some international support.
1950	China and the Soviet Union sign a treaty of friendship. Large amounts of aid and technical assistance are provided to help China's development.
	Chinese forces enter the Korean war in support of North Korea against United Nations forces.
1951	USA agrees to provide Taiwan with military equipment in defence against possible attack by the PRC.
1954	USA signs a mutual defence treaty with the Republic of China (Taiwan). There is conflict in the Taiwan Straights as the People's Republic bombards the offshore islands of Quemoy and Matsu, held by the Guomindang.
1959	Rebellion in Tibet crushed by PLA; the Dalai Lama and refugees flee to India.
1960	Soviet Union withdraws its aid after bitter ideological disputes, though relations ostensibly remain friendly.
1962	Chinese troops defeat India in battles on the Himalayan borders, and long period of animosity begins.
1964	China successfully tests a nuclear bomb.
1969	Chinese and Soviet troops in battles at the border on the Ussuri River.
1971	The PRC is admitted to the United Nations as the legitimate government of China, beginning a process of increasing international participation for it and reduced recognition of the Taiwan regime.
1972	US President Nixon visits Beijing and ends American isolation of the PRC, starting a process of closer economic and diplomatic relations which shift the power balance of the world.
1979	China involved in war with Vietnam along their border to put pressure on Vietnam to end its interference in Kampuchea; Vietnam, in effect, defeats China.
1984	Britain and China sign an agreement for the return of Hong Kong to the PRC in 1997.
1987	China and the Soviet Union agree in principle on how to solve their border dispute, reflecting improvements in their relations.
	Demonstrations in December in Tibet for independence from Chinese rule lead to clashes, with dead on both sides.
1988	Riots in Lhasa in March and December after demonstrations for greater autonomy and human rights.
	Naval clashes between China and Vietnam over control of the Spratley islands. Gorbachev pledges that Soviet troops will be steadily withdrawn from the Mongolian People's Republic.
1989	India's Prime Minister, Rajiv Gandhi, visits China, the first Indian premier to do so for thirty-four years. Agreement to set up a joint working group to seek a border agreement.
	Vietnam's deputy Foreign Minister visits Beijing to discuss disputed issues, especially Kampuchea/Cambodia.
	June: Gorbachev visits Beijing (during unprecedented mass demonstrations for democratic reform and an end to corruption) for the first meeting between Soviet and Chinese government leaders for thirty years. He announces further troop withdrawals from Mongolia and along the Sino–Soviet border.

GEOGRAPHIC FACTORS SHAPING SECURITY

One analyst of military strategy points out that 'defending China means defending virtually an entire continent' (Segal 1985: 11). Evidently its vast land area – it is the third largest country in the world – poses significant problems to China's military planners. This territory has to be defended while exercising care about relations with twelve bordering states. China has the longest land boundaries of any state, while to the east and the south it has a long coastline. Given these aspects of its geography, it might be expected that China would have developed strong land- and sea-based defences. But a more accurate view of events needs to guard against seeing geographical factors as deterministic. At various times China's leaders have chosen to neglect its sea- or land-based defences. Indeed, historically the preoccupation has been with the threat from the north, from the highly mobile inhabitants of the territories of Inner Asia, fear of whom is so powerfully shown by the Great Wall. For much of its history, sea defences have had little priority. However, since 1949 Chinese military planners have developed strong land- and sea-based forces.

The distribution of population, which is highly concentrated in the eastern half of the country, means that in the event of conflict with the Soviet Union to the north and west, Chinese forces could retreat and trade space for time. However, cities such as Beijing and Tianjin and the strategically vital industries of China's northeast (Manchuria) are relatively close to China's northern borders and are vulnerable, especially in circumstances of alliance between the USSR and People's Republic of Mongolia. Important coastal cities such as Shanghai and Guanzhou are very vulnerable to attack from the sea.

China's ethnic geography presents its leaders with particular strategic problems and opportunities. Although 94 per cent of the population are Han Chinese, the minority peoples largely occupy the territory outside China's 'core' area of the densely settled east. Hence many border territories are peopled by minority nationalities. The danger is that such groups might wish to break away from the Chinese state and that foreign powers might be tempted to encourage such rebellions. Conversely, as minority peoples often straddle the boundary (e.g. the Uygurs in Chinese Xinjiang and Soviet Central Asia), China could also adopt the tactic of encouraging minority grievances against a neighbouring power.

Are such geopolitical issues still relevant in a nuclear age? Clearly they are, given that all military conflicts except one have been non-nuclear. The geopolitical factors described above have been one element in shaping all China's conflicts since 1949. For example, there is some evidence that in conflicts with the Soviet Union in 1962 and 1969 both states tried to encourage minority grievances. There is also evidence that they both limited these conflicts, seeing that minority unrest could threaten control of 'their' minority peoples.

Mention of nuclear issues brings out one of the dangers of a geopolitical analysis – that of assuming a static 'geography' shaping state policy. The internal geography of the country changes: for example, Tibet's connections with the rest of China have been greatly altered by new roads built since 1959. So has China's geostrategic position with respect to other states. The development of long-range nuclear bombers and intercontinental missiles means that China is both threatened by the nuclear powers and can threaten other countries. Thus China's effective position with respect to other states has been changed significantly by nuclear weapons.

Other, often longer-term, changes have also transformed China's leader's perception of its location. Historically, China was relatively isolated from other major powers and civilisations. It saw itself as Middle Kingdom (Zhong Guo), central to the world. But from the eighteenth century China had to confront other states which were culturally and militarily powerful. Since then, its leaders have had to operate on the basis that their country is, like others, subject to threats from near and far.

Indeed, in the twentieth century, China has been located geostrategically between Russia as a developing power in Asia, and the United States emerging as a Pacific power. Moreover, Japan – once culturally subordinate to China – emerged to colonise and invade China until 1945. Thus without even considering the impact of nuclear weapons, China's view of its place in the world has profoundly changed. Later in the chapter we will return to these wider questions. First, let us consider the extent of China's legitimate claims to territory, as seen by the post-1949 leadership. This issue remains crucial to an analysis of China in the 1990s.

DEFINING AND DEFENDING THE CHINESE TERRITORY

The decline of the Chinese state's ability to govern effectively in the nineteenth and first half of the twentieth century led to the growth of autonomy in areas more distant from the core of the country, and encroachment on China's territory by foreign powers. From the eighteenth century, at a time of domestic disintegration, China came into increasing contact and conflict with militarily powerful states, the most recent of which was Japan. In addition, the civil war struggle perpetuated a division of the country: the continued separation of Taiwan from mainland government (it was a Japanese colony from 1895 to 1945). These issues are central to an analysis of the CPC's outlook and its policies in post-1949 China, in which maintaining territorial integrity has been of great importance.

The successful communist seizure of power in 1949 can in part be explained as a nationalist reaction to Japanese aggression. In that year, Mao Zedong said that the Chinese people, had 'begun to stand up ...and that our national defence would be consolidated and that no imperialist would be allowed to invade our territory again' (Schram 1969: 167–8). Such nationalist feelings possibly included hopes of regaining vast territories once controlled by China. In 1936 Mao Zedong had asserted that, when the communist revolution succeeded, the 'Outer Mongolian Republic will automatically become a part of the Chinese federation, at its own will. The Mohammedan and Tibetan peoples, likewise will form autonomous republics attached to the Chinese federation' (Schram 1969: 419–20). In 1939 Mao wrote:

> After having inflicted military defeats on China, the imperialist countries forcibly took from her a large number of states tributary to China, as well as part of her own territory. Japan appropriated Korea, Taiwan, the Ryuku islands, the Pescadores and Port Arthur; England took Burma, Bhutan, Nepal, and Hong Kong; France seized Annam; even a miser-able little country like Portugal took Macao from us.
>
> (Schram 1969 p. 375).

Great care is needed in using these statements to analyse Chinese policies after 1949 and into the 1990s. They may reflect widely held nationalist feelings, but political leaders have to confront the realities of power. For example, although many Chinese leaders saw Outer Mongolia as being within China's sphere of influence, in 1950 China recognised the Mongolian People's Republic. Conceptions of what is legitimately China's territory (or even sphere of influence) may change, and the leaders in the 1990s may not have the same conceptions as did Mao Zedong's and Deng Xiaoping's generation. The leaders of the revolution probably found it difficult to disentangle whether it was the loss of territory that was resented, or rather the fact that foreign powers could do with China as they wished.

What territories under other states' control have been seen by the Chinese as legitimately part of the Chinese state since 1949? The issue of 'Outer Mongolia' has been passed over, given the major influence of the Soviet Union in that region, though there is evidence that some CPC leaders were opposed to recognising the Mongolian People's Republic. (The term 'Outer Mongolia' and 'Inner Mongolia' evidently reflect a Chinese view of those territories' positions.) The most crucial issues since 1949, which will continue to be of great significance, can be divided into two.

The first concerns territories which are claimed as part of China by the PRC, and whose inhabitants generally accept that their territory forms part of China, though most might not want to be governed by the PRC. With these, the argument is about foreign control (Hong Kong and Macau, and in the past Manchuria), or which government is the legitimate one (Taiwan). This category, can be referred to as *irredentism*: the PRC wants to restore to its control areas inhabited by Chinese people which have been separated from their state.

The second relates to areas of non-Han Chinese nationalities, which are currently under Chinese control, where there are significant numbers of people who reject the Chinese state and argue for greater autonomy or actual independence (secession). Of these, Tibet is the most famous, though similar problems exist in Xinjiang and Inner Mongolia. These are discussed as *nationalist opposition*.

Irredentism

The PRC leaders' determination to re-integrate Taiwan into China is an example of a basic geopolitical continuity between pre-1976 and post-Maoist China. Continuity is also evident if one considers PRC policy towards Hong Kong and Macao. To Mao and other Chinese leaders these territories (including Taiwan) were seized by foreign powers through a series of what the Chinese have termed 'unequal treaties'. This term is applied to a variety of territorial issues (see the following discussion of border disputes). The treaties were unequal in that, although the Chinese government may have formally agreed them, it did so as a result of being attacked or threatened with force.

Though not a current concern, the CPC inherited a similar problem of foreign control over former treaty territory in Manchuria. This region came under Russian influence in the latter part of the nineteenth century as a result of Russia's eastward expansion and subsequent enforced treaty claims. It passed into Japanese influence after 1905, and became a Japanese colony until Japan's defeat in 1945. The late entry into the war against Japan by the Soviet Union enabled the Soviets to claim and take away large amounts of industrial equipment from the northeast, and in effect reassert control in what it seemed to see as its old sphere of influence.

The 1950 treaty and alliance which the Soviet Union signed with the new People's Republic of China recognised Manchuria as an integral part of China, though even then the Soviets retained control over some railways and naval bases. China's entry into the Korean War (1950–3) on the side of North Korea is partly explained by the need to secure its control over the region of Manchuria, which is an extremely important base of heavy industry.

Taiwan

This large island off the south coast had been seized and colonised by Japan in 1895. During the Second World War both the Cairo and the Potsdam declarations had stipulated that all territory which Japan had taken from China, including Taiwan (Formosa) should be returned; on the defeat of Japan in 1945, it was reincorporated. In 1949, members of the Nationalist (Guomingdang) Party and many of its troops fled to Taiwan, after their defeat by the communist armies in the civil war. In 1950 China intervened in the Korean war on the side of the north, to resist the growth of American power in the region.

President Truman ordered the US Seventh Fleet to patrol the Taiwan Straits, effectively preventing Taiwan's integration into the new Peoples' Republic of China (PRC). In 1951 the USA signed a mutual defence agreement with the nationalists in which the USA undertook to provide them with 'military material for the defence of Taiwan against possible attack'. Thereafter the USA continued to build up the nationalist forces and to recognise their government in Taiwan ('The Republic of China') as the legal government of all China.

As America's power in the Pacific declined following its defeat in Vietnam, the US government sought to improve relations with the PRC. The process began in the latter years of the Vietnam war, marked by President Nixon's visit to Beijing. There followed the establishment of full diplomatic relations in 1979, and the Sino–US communiqué on Taiwan in 1982. With these the USA recognised the PRC as the legal government of China and that Taiwan is part of China.

However in 1979, by a separate agreement with Taiwan, the USA declared its policy of 'preserving and promoting extensive, close and friendly ... relations between the peoples of the United States and Taiwan'; it expected 'that the future of Taiwan will be determined by peaceful means' and that the USA would continue to provide arms of a defensive character. Indeed, given the island's relative prosperity compared with the PRC and its military and diplomatic ties with the US, its re-integration seems less likely than it did in 1949, although there are now growing trade and cultural links between the mainland and Taiwan as part of Beijing's policy of promoting peaceful reunification.

Hong Kong and Macao

Until 1976 Chinese statements on Hong Kong and Macao emphasised that foreign control was the result of unequal treaties and that China would settle these questions when the time was ripe. For instance, in 1972 China's representatives at the UN stated 'the question of Hong Kong and Macao belongs to the question of unequal treaties which the imperialists imposed on China'.

In spite of this, both of these 'parts of China temporarily under foreign control' have proved of commercial value to the PRC. Hong Kong, especially, provides significant economic advantages, since for many years the PRC was not diplomatically recognised by many powers (particularly the USA), and access to foreign technology and goods was difficult. China's exports to and through Hong Kong brought in essential foreign currency to buy technology from abroad.

However, a major part of the colony's land is actually on a lease which expires in 1997, and, as the expiry date approaches, Chinese leaders have been prompted to work for the return of both Hong Kong and Macao. Between 1982 and 1984 the Chinese and British governments negotiated

Hong Kong's future. In September 1984 they signed a declaration which stated that China 'upholding national unity and territorial integrity ... [had] decided to resume the exercise of sovereignty over Hong Kong' in July 1997. In March 1987 China and Portugal signed a similar agreement by which Macao was to revert to Chinese sovereignty in December 1999.

The present government in Beijing is willing to jeopardise Hong Kong's economic advantages to China by taking it back in 1997, but will try to maintain Hong Kong's prosperity and even its ostentatious capitalism. The 1984 declaration provides that Hong Kong will become a 'Special Administrative Region' and enjoy a 'high degree of autonomy ... except in foreign and defence affairs ... [maintaining its] current economic and social systems'. These arrangements, which essentially guarantee Hong Kong's capitalist economic system, were to 'remain unchanged for fifty years'. (Similar guarantees were subsequently given for Macao.)

The agreements over Hong Kong and Macao need to be seen in the context of China's opening to the West and the creation of Special Economic Zones and open port cities (see Chapter nine). The PRC also sees these agreements as offering a formula that could permit Taiwan's reintegration with the mainland. After signing the Hong Kong agreement the Chinese government stated that 'the Taiwan issue can also be settled by the method of "one country, two systems", and on more liberal terms for, among other things, Taiwan would be allowed to keep its armed forces' (*Beijing Review* 3 February 1986).

Nationalist opposition

China's communist leadership not only inherited territory which was still basically that of the Qing dynasty Empire, it also retained much of the thinking that went with Han-Chinese views of control over it. Instead of recognising that areas inhabited by other nationalities were conquered or colonised territory, the CPC continued to see them as inalienable and inherent parts of the People's Republic. As a result, pre-existing resistance to Chinese rule has not abated in areas where other nationalities live. Indeed, the policies of the party in repressing other cultures and religions have, if anything, provoked more hostility from many minority peoples. These issues are considere in more detail later.

Xinjiang and Tibet

For much of the first half of the twentieth century these regions were outside the control of the Chinese government. Manchuria came under Japanese influence after 1895, and the two far-west territories were a low priority given the collapse of Imperial power. The post-1949 government was determined to reintegrate them into the Chinese state.

This policy emerged during the revolution, and in 1936 Mao had asserted that the Muslim and Tibetan peoples would become part of China. For some of them this transfer was achieved by units of the People's Liberation Army (PLA) moving west and southwest after the core areas of east China had been secured.

Xinjiang had effectively stayed out of Chinese government control after the 1911 Republican rule disintegrated, and a succession of warlords became the effective rulers (though the Soviet Union also intervened for a time). One way that its reintegration was achieved was by large numbers of troops (both PLA and Guomindang) being demobilised in Xinjiang from the 1950s onwards and being forced to settle there and develop farming and industry.

Tibet posed a more difficult military problem, given the terrain and distances involved. Moreover, Chinese control had not been very extensive or solid, even compared with that in Xinjiang, and Tibet enjoyed considerable *de facto* autonomy. But whatever the Tibetans wished in the past or now, China's leaders both communist and nationalist have always asserted that Tibet is a part of China. In 1959, the PLA defeated a Tibetan uprising which involved armed rebels, in order, as Zhou Enlai put it, to 'liberate the Tibetan people and defend the frontiers of China'.

Manchuria, Tibet and Xinjiang thus represent territories which had effectively gone out of central control but which the post-1949 government has succeeded in reintegrating into the Chinese state. In addition, there are the territories of Hong Kong and Macao, which passed into foreign control in the nineteenth century, and Taiwan, which has been separated as a result of the revolution, for which the PRC maintains its irredentist claims.

DEFINING AND DEFENDING THE BORDERS

Previous discussion has emphasised Chinese sensitivity over 'unequal treaties' and concern to establish its territorial integrity. The same issues come to the fore when analysing post-1949 boundary policy. The PRC inherited the border problems of previous Chinese regimes. The European concept of permanent boundaries, accurately delimited by treaty and then demarcated on the ground came late to China. Traditionally, the Empire operated a policy of fluctuating territorial control, with different areas moving in and out of Chinese administration and influence. Frontier zones existed on the periphery of influence, but fixed boundaries were not established. As China came into contact with powerful European states so its boundaries began to be established. For instance, those with the Russian Empire were defined by a series of treaties dating from that of Nerchinsk in 1689. By these treaties vast areas of what is now the Soviet Far East and Soviet Central Asia passed out of Chinese control and sovereignty.

The PRC has adopted a distinctive policy towards its inherited boundaries. With the exception of those territories regarded by the Chinese government as legally Chinese (e.g. Tibet, Xinjiang, Hong Kong and Taiwan), China declared its willingness to respect the boundaries established by previous treaties. For example, government statements about the Sino–Soviet border (even at the height of military clashes there in 1962 and 1969) made explicit that China was not asking for the return of lost territories. Post-1949 Chinese governments have adopted a consistent policy to boundaries where the line is in dispute or where the process of delimitation and demarcation is incomplete. They have insisted that the whole boundary line should be established by a new (equal) treaty.

China has not accepted the principle suggested by the Soviet Union and others that negotiations should consider just those sections where previous treaties and/or maps have been imprecise. The Chinese position is that the treaties that established these boundaries are 'unequal' ones, forced on a humiliated China; new, equal treaties are necessary to delimit a freely accepted boundary. There were particular problems about many of China's boundaries, since many of the treaties were imprecise and few had been marked out on the ground.

As the PRC established diplomatic relations with neighbouring states, so it concluded a succession of border agreements with those states (e.g. Burma and Nepal 1960; Mongolia 1962; Pakistan and Afghanistan 1963). With these, China's attitude to boundary negotiations was accepted. However, since 1949 China has been in dispute concerning its borders with the Soviet Union, India and Vietnam. Up to mid-1989 they remain disputed, though there have been recent moves towards accommodation.

In analysing these conflicts from a geopolitical perspective, it is necessary to disentangle the extent to which they are about the territory and boundaries as such, or much wider issues of state-to-state relations which have led to overt conflict over disputed boundaries.

The Sino–Soviet boundary

The border with the Soviet Union is the longest land boundary in the world but much of it runs through inhospitable, economically unimportant territory. Some of the maps are imprecise and there are disagreements between the USSR and the PRC as to the boundary in the Pamirs and the precise line along the Amur and Ussuri rivers in China's northeast (Figure 11.1). Where parts of the Amur and Ussuri rivers form the boundary the valleys are relatively fertile, and the border runs through some areas which are strategically vital. In 1962 there were military conflicts in the central Asian section of the boundary. These were relatively small-scale and at the time neither side publicised them. Ostensibly, this was still a period of co-operation and friendship between two socialist states.

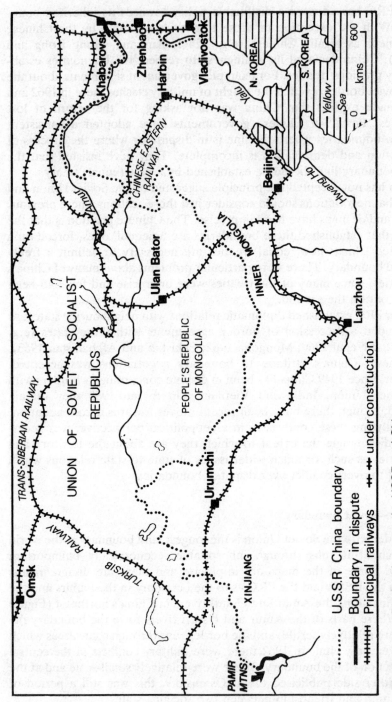

Figure 11.1 Borders between the USSR and China

After 1963 the two states were in open diplomatic conflict over a wide variety of issues. To explain these conflicts would take us into how the two states saw each other's domestic policies, and the complex interactions of the military and diplomatic relations of the power triangle of the Soviet Union, USA and China (discussed later). The issues that drove the two states into conflict had little to do with their boundary *per se*, but their effect was to lead in 1969 to large-scale conflicts on and near the boundaries. In particular, they focused on an island (Damansky or Chenpao) on the Ussuri River. Thereafter, relations between the two states deteriorated, though on the borders there were only small-scale 'incidents' in the period *c*. 1969–80.

The threat was real enough: the Soviet Union deployed about a quarter of its troops in the Soviet Far East and the Mongolian Peoples' Republic, clearly aimed at China. Included were SS20 missiles within easy range of China's northern cities and industrial regions. About half of China's forces were positioned to defend its territory against possible Soviet attack. The impact of this threat on China's military strategy and development was immense. Having moved public and strategic facilities inland to avoid the Japanese in the 1930s, and away from the coast and the US threat in the 1950s, a repeat of these strategies occurred from about 1965 up to 1972, but this time to escape possible Soviet invasion (see Chapter two). This *sanxian*, or Third Front policy, entailed massive investments in new factories and research institutes in the centre and southwest of China (see Figure 2.4 on p. 38).

The immediate result of China's post-1976 policies was a widening of the dispute with the USSR. The Soviet Union initially attacked China's domestic policies for moving away from socialism. More significantly, the Soviet Union feared China's increased links with the United States – there was even talk of China being an unofficial member of NATO. China, meanwhile, was in conflict with Vietnam (see pp. 280–3) and resented the Soviet Union's military and diplomatic links with Vietnam and its support for Vietnam's occupation of Kampuchea. The effect of these issues was to reduce prospects of settling the Sino–Soviet border question.

However, recently there has been a distinct improvement in Sino–Soviet relations, progress on border relations and a relative absence of border incidents. In July 1986 Mikhail Gorbachev gave a major speech on PRC–USSR relations. Significantly, it was in Vladivostok, a Soviet naval and military base and fulcrum of its power in the Far East. He announced that talks had begun with the Mongolian People's Republic on the withdrawal of Soviet troops, and this was begun in 1988. Gorbachev also recognised the Chinese claim that the border ran down the central river channel, thus accepting the Chinese claim to Damansky island. In August 1987 discussions ended with agreement in principle on how to settle the boundary in this eastern sector. Soviet and Chinese officials will now attempt to establish the detailed line in this sector and then proceed to negotiate over the western sector. If progress is satisfactory a new Sino–Soviet treaty will then delimit the whole of the boundary line. The post-1976 Chinese government will then have achieved the negotiating position asserted by Mao's administration.

The Sino–Indian boundary

China's other long-standing boundary dispute is with India. At one level this dispute can be seen as a result of both states' determination to act as a regional power in Asia. Despite a common history in struggle against imperialism and sympathy for each others' concerns in the 1950s, there were areas of dispute. Tibet was and is regarded by China as an integral part of the Chinese state, while historically India had cultural links with it.

But the major issue has been conflict over the Sino–Indian boundary. The border was largely formed by Britain as it advanced northwards to protect the British Raj (partly against a perceived threat from Tsarist Russia). At Simla in 1914, McMahon, a British official, persuaded Tibetan officials to agree to a boundary line in the eastern sector to the east of Bhutan. This McMahon line moved British control substantially northwards. No Chinese government before 1949 accepted this line.

After 1949 China and India differed, in that Chinese maps showed the boundary (particularly along the McMahon line) as not legally delimited. There were also major differences in the territories which both sides claimed, including large areas of India's Northeast Frontier Region (Figure 11.2). There were also disagreements about the central section west of Nepal, and the far west in Aksai Chin, where China disputes an area larger than Belgium. This section is stategically vital to China, and in the late 1950s China built a (military) road here linking Tibet to Xinjiang. The first that India knew of this road (through what it regarded as its territory) was its appearance on a Chinese map!

In 1962 India decided on a forward military strategy to assert control over the disputed territories, particularly in the eastern sector. China charged India with aggression, and Chinese troops advanced to occupy the disputed territories and defeated India. These events still shape contemporary Sino–Indian boundary relations. Border talks did commence in 1981, and by 1989 progress had been made, boosted by Indian premier Rajiv Gandhi's visit to Beijing. There have been troop withdrawals on both sides of the border, and a compromise is in sight. Essentially this entails India recognising China's claim to Aksai Chin in the west and for China to recognise India's claims in the east – accepting the McMahon line! This is not a change in China's policy. In 1956 Zhou Enlai told Nehru that China objected to the name of the MacMahon line but the boundary itself could be broadly acceptable to China. Indeed the 1962 Sino–Burmese

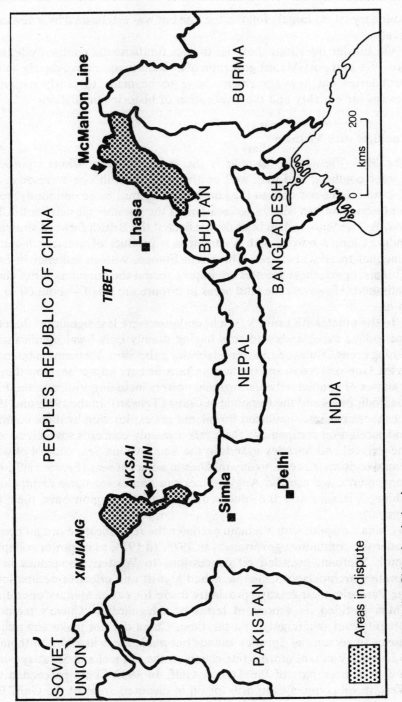

Figure 11.2 India–China borderlands

boundary treaty largely follows the line but was established by a new equal treaty.

On border questions there is, thus, a fundamental continuity between pre-1976 and post-Maoist governments. China is to achieve clearly defined boundaries but new agreements have to be made to satisfy nationalist desires for equality and the eradication of historic humiliations.

Conflicts with Vietnam

The Sino–Vietnamese dispute is the most recent of China's post-1949 border conflicts, and may well be the most difficult one to resolve. The dispute relates not only to the land boundary but, more ominously, to the confused claims to islands and sea-bed in the possibly oil-rich South China Sea. As the Sino–Indian boundary reflected the British forward strategy in India, China's border with Vietnam is a product of nineteenth-century 'unequal treaties' between China and France, with its colonies in Indo–China. In part, these treaties are imprecise and the boundary is not clearly delimited. However, the land areas in dispute are small – about 60 sq. km in all.

In the nineteenth century, sea boundaries were less significant than they are today, the islands and seas having mainly only local significance as fishing areas. Thus the treaties establishing the Sino–Vietnamese boundary in the Gulf of Tonkin and the South China Sea are imprecise. Now they are a matter of conflict between regional powers including Vietnam, the PRC, the Philippines and the Republic of China (Taiwan). In the 1970s and 1980s these seas became significant for oil and gas exploration by these countries and foreign oil companies. The conflict mainly concerns sovereignty over the Paracel and Spratley islands in the South China Sea, control of which would legitimate access to any oil finds in adjacent seas (Figure 11.3). (For convenience we use the Anglo–American names for these archipelagos: China, Vietnam and the other countries of the region have their own names.)

China's dispute with Vietnam predates the re-unification of that country under the communist government in 1975. In 1973, as its power collapsed, South Vietnam awarded oil concessions to Western companies in the Spratley archipelago. China launched a swift and effective occupation of the Paracels: their relative proximity made for easier military operations. China justified its actions in terms of safeguarding 'China's territorial integrity and sovereignty'. At the time, China did not make any military moves to secure the Spratley islands but made clear its claim to them.

In 1982 Vietnam affirmed its claims to the Paracel and Spratley islands and to a large part of the Tonkin Gulf. In 1983 China proceeded with Western oil companies to drill for oil in disputed areas of the Gulf. Both sides increased their naval patrols, and China established what it termed

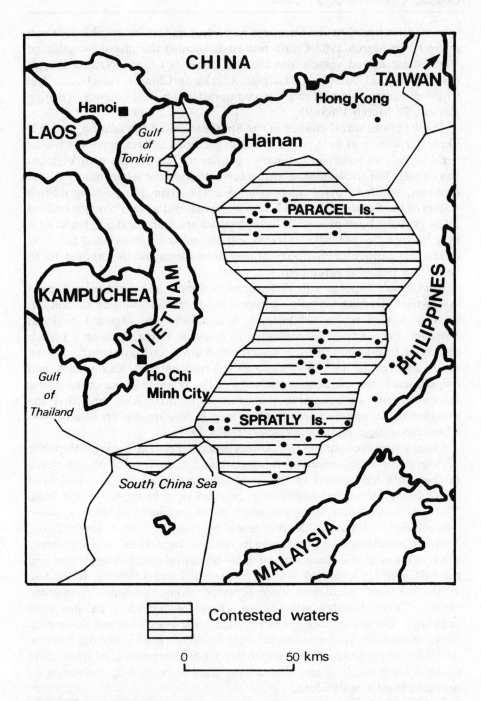

Figure 11.3 Conflicts over sea boundaries
Source: Chang, Rhee and Macaulay

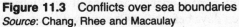
Contested waters

0   50 kms

'observation posts' in the Spratleys as part of an 'oceanographic research project'. In March 1988 China protested 'against the illegal intrusion of Vietnamese armed vessels into the sea waters of China's Nansha Islands and the flagrant launching of armed attacks on Chinese vessels. ... The Chinese vessels were forced to counterattack in self-defence' (*Beijing Review*, 28 March 1988: 9).

These serious naval clashes in the Spratleys raised the tension, and it is likely to increase as both try to establish garrisons and reinforce the fragile coral islands for naval and military equipment. The dispute with Vietnam has already led to warfare, and this tends to hide the wider nature of the problem, which involves other countries who claim the possibly oil-rich waters of the South China and East China seas. But so far no major finds of oil or gas have been proven. If large deposits are found and oil prices rise it may be that the potential rewards and the oil companies need for clear claims and a peaceful environment may drive China and the other states to come to a peaceful agreement.

But China's disputes with Vietnam are not solely about either land or sea boundaries. Despite China's significant aid to the Vietnamese revolution, there is now a revival of historic, nationalistic and regional rivalries. Vietnam, once a state subordinated to Imperial China, became a French colony and, re-united in 1975, sees itself as a regional power in Southeast Asia. Indeed, in 1977–8 Vietnamese forces entered Kampuchea and helped defeat the Chinese-supported Pol Pot regime. China charged that Hanoi 'dreams of becoming the overlord in S.E. Asia' and that the invasion of Kampuchea was the first step in 'the rigging up of an Indo–China federation' (*Beijing Review*, 21 July 1978).

Other issues brought the two powers into conflict. The Socialist Republic of Vietnam's collectivisation campaigns and state control of commerce in the south led to tens of thousands of ethnic Chinese who had lived in Vietnam for generations being expelled or seeking to escape from Vietnam. China's sensitivity and anger at the treatment of fellow Chinese was coupled with the problems of resettling many of them in south China. Vietnam maintained and increased its ties with the Soviet Union, granting it naval bases at the same time as Sino–Soviet relations deteriorated and the PRC and USA moved closer diplomatically and militarily. In the late 1970s increased 'incidents' were reported along the Sino–Vietnamese border. These disputes often began where the boundary treaties were imprecise. But such differences over land boundaries were not the central cause of conflict. Wider questions were involved, principally the Chinese desire to put pressure on Vietnam to get out of Kampuchea, which it could do by forcing much of the Vietnamese army to be tied up defending its northern border with China.

In 1979 Deng Xiaoping's government launched an assault on Vietnam which, though it lasted only sixteen days before the Chinese withdrew, was a major war. For China it was the largest conflict since the Korean war. Though both states had heavy casualties, China was, in effect, defeated. Its poorly equipped troops were no match for the well-led, experienced Vietnamese forces. Vietnam continued to help establish a regime friendly to it in Kampuchea and to increase its connections with the Soviet Union.

Since then there has been little progess in resolving the Sino–Vietnamese dispute. Negotiations have furthered the split by emphasising disagreements over the boundary in the Gulf of Tonkin. There were signs in early 1989 that the Vietnamese withdrawal from Kampuchea would encourage China to adopt a more conciliatory attitude, and the continued reduction in tension with the USSR will also help by removing Chinese fears of Soviet collusion with Vietnam.

CHINA: A UNITARY MULTINATIONAL STATE

While engaged in defining and defending its territory with respect to other states, the Chinese government has effectively had to integrate the state area, that is to bind together the peoples and the territory of what it considers to be China. Analysis of how this has been attempted would require examinination of the organisation of government and the role of the party and the army, the role and organisation of the media and education, and much more. Here I examine only the more immediately geographical issues raised in the introduction. I mention briefly the problems of integrating a large state area and then consider in more detail the problem of integrating the areas where minority peoples are significant.

These were two of the central problems Mao Zedong emphasised in 1956 in a major secret speech to party officials. The problem of reconciling regional interests with the state's concern for unity is a problem for all states. China's size and poor communications, and its recent history of regional warlord powers, posed further difficulties for achieving 'national' unity. Mao's 1956 speech emphasised that to enable regional development 'you must arouse the enthusiasm of the regions . . .; the regions must have an appropriate degree of power'. However Mao also emphasised 'we want unity . . . the independence sanctioned by the centre must be a proper degree of independence. It cannot be called separatism.' (Schram 1974: 71–3).

The danger of separatism, of an area or people threatening to break away from the state, is posed most acutely by the relationship between the Han Chinese nationality and the other peoples (national minorities). The traditional Chinese state was a land empire bringing into its control non-Han cultures with their own economies and political systems. China's dynastic governments had seen these non-Han peoples as part of a Greater China. Indeed, beyond those conquered or incorporated peoples they regarded vast areas of Asia and their different states (e.g. Korea, Outer

Mongolia, Nepal and Vietnam) as subordinate to the Chinese Empire even if not part of it.

By 1949 many previously distinct peoples had long since been assimilated into Han culture, such that 94 per cent of the population of the state was ethnically Han. But the geographical distribution of the ethnic minorities posed a major challenge to any Chinese government. For, though some of the minorities were dispersed in what were now essentially Han areas, certain ethnic groups were concentrated in places where they formed a clear majority of the population. They were the majority in approximately 50 per cent of the Chinese territory to the north, southwest and west of 'China proper' (see Chapter four). These areas today are, for the most part, included in the Inner Mongolian, Xinjiang Uygur, Tibetan, Ningxia Hui and Guangxi Zhuang Autonomous Regions and Yunnan province.

Before 1949 the Chinese Communist Party had considered how to deal with minority issues. In 1936 Mao Zedong had seen these areas becoming part of the Chinese federation at 'their own will'. At times the party had even suggested recognising the right of secession. However the PRC's first Constitution in 1954 stipulated that China is 'a unitary multinational state' of which the 'national autonomous areas are inalienable parts'. Although opting for a unitary state ruled out a federal structure, certain quasi-federal features have developed. These are most obvious in the relatively powerful provincial authorities of Han-populated areas.

However, the regional autonomy which minorities have supposedly been granted with their special status has been sharply circumscribed by the central exercise of party and army power. The army and party posts in the Autonomous Regions have been held by Han Chinese, though recently matters have changed slightly. Large numbers of (mainly Han) troops have been stationed in the border areas, particularly in the early years of establishing control and wherever local revolt (as in Tibet in 1959 or 1987) has threatened control from the 'centre'.

Government statements have been quite explicit in emphasising the economic value of these territories. For example, Mao's 1956 speech said 'While the Han nationality has a large population, the national minority areas have riches under the soil which are needed for building socialism' (Schram 1974: 74). Official Chinese statements also emphasise the aid these areas have received in the development of their regional economies. Certainly, central government funds have been and still are channelled towards these regions (see Chapter two) to develop road and rail links, to establish industry, modernise agriculture and develop education and medical services. But there is dispute over how much of the benefit is gained by the local people and how much is for strategic purposes or even to bolster the large-scale land colonisation by Han peoples in, for instance, Inner Mongolia and Xinjiang.

Care has to be exercised in using Western analyses (including this one) of China's policies towards these border areas. Some writings have a very romantic view of the minority peoples' cultures. Views of Tibet as a Shangri-la, and of pastoralist peoples as having some intrinsic superiority, shape some analyses. However, other Western sources can be blind to what has been destroyed since 1949 in the name of socialism, and fail to ask who has decided that this 'development' is beneficial, what form it should take and who should benefit.

Mao's 1956 speech emphasised the danger of Han chauvinism. Contemporary Chinese statements now recognise that much of the practice of policy since 1949 has been at the expense of the area's land and peoples. At times, policies have been overtly aimed at those people's assimilation, denying their language, culture and religion. Development policies have been imposed from outside, failing to recognise distinctive local economies and ecological conditions, as in the frequent attempts to destroy nomadism or the Maoist exhortations to grow grain everywhere (see Chapter six). From the perspective of the 'centre', of the dominating Han culture, there is a logic in encouraging Han migration to the minority areas, for these migrants will have skills which will lead to the area's economic development. However, from the indigenous peoples' perspective this migration may be less than welcome.

Since 1978 there has been more open discussion of the problems of developing policies towards these areas. Government statements give much greater emphasis to training minority cadres, to respecting minority cultures and to economic development more suited to local conditions. Spending on education and health care has been increased and no taxes are to be levied until 1990. Minority families are not restricted to having one child. In effect, there is encouragement by the centre for minority peoples to increase their overall share of the population (see Chapter five).

Thus there have been significant changes since 1978 in official recognition of the problems of Han relations with the minorities and moves to develop a more equal relationship. It remains to be seen whether these promised reforms will be effected and the basic geopolitical problem of the distribution of the minority peoples 'solved'. The recent experience of Tibet suggests the problem is not going away, and at least some of the inhabitants prefer much greater autonomy if not outright independence. This region is dealt with individually here, though it is as well to keep in mind that the issues involved apply to many other minority peoples in China which receive less media attention in the West.

Tibet

Tibet was briefly conquered by the Mongols, who established their rule over China in the Yuan dynasty (AD 1279–1368), and the region came under Manchu control in the eighteenth century. Both of these 'foreign' rulers of the Chinese Empire followed Tibetan Buddhism, the Mongols having brought it back to their homeland during the Yuan. The extent and nature of Chinese influence and control over Tibet is disputed, and it is ironic that in the periods when Tibet was most strongly linked to China, the Chinese themselves were under foreign rulers. In the first half of the twentieth century, Tibet 'enjoyed' *de facto* independence. But whatever the Tibetans wanted in the twentieth century, Chinese leaders have asserted that Tibet is Chinese territory.

To the British ruling India, Tibet was a buffer state in which Britain was willing to intervene to protect British power. Particularly odious to the Chinese (nationalist and communist alike) was the Simla conference of 1914 at which representatives of Britain, Tibet and China recognised Chinese suzerainty (i.e. control of foreign relations), but not sovereignty, over Tibet. No Chinese government has recognised that agreement, but the turmoil and disruption of civil war and Japanese invasion meant that until the CPC came to power there was very little Chinese involvement in Tibet. Likewise, there was no Tibetan involvement in the Chinese communist victory, and if the new regime was to impose its control it meant bringing in revolution from outside.

In 1950, the PLA clashed with Tibetan troops as the new government sought to re-integrate Tibet into the Chinese state. Then, in 1954, China and India reached an agreement on Tibet, by which India recognised it to be an integral part of China, and China undertook to respect religious and cultural traditions. But in 1959 a rebellion against Chinese rule broke out in Tibet and the Dalai Lama and many of his followers fled to India, where they still remain. China has seen India's hosting of the refugees as an obstacle to improved Sino–Indian relations.

The Chinese government increased the numbers of troops in Tibet to ensure its 'security'. Tibetans charge that inappropriate policies have been imposed on the region, and religious and cultural activity suppressed. In 1980 the then Communist Party General Secretary, Hu Yaobang, made an inspection tour of Tibet, after which new, more sensitive policies for Tibet were announced. The adoption of the family responsibility system in agriculture was encouraged. This has 'contributed to the revival of traditional animal husbandry – the key sector in the Tibetan economy . . . large tracts of pasture land formerly destroyed to grow grain have again been sown with grass for foraging' (*Beijing Review*, 15 October 1984).

That China felt able to allow Western tourists into Tibet from the early 1980s indicated a certain confidence about its control. However, on a number of occasions since, especially in 1987 and 1988, there were demonstrations in Lhasa and elsewhere against Chinese domination. These were suppressed, resulting in the death of significant numbers of Tibetans and some Chinese police. Beijing reasserted that 'Tibet is an inalienable part of Chinese territory' (*Beijing Review*, 19 October 1987: 14). The concessions to the grievances of Tibetans and other minority peoples have clearly been insufficient to satisfy many of them. The Chinese government response, in a major speech on the nationality issue by the then Party general secretary Zhao Ziyang, has been to recognise the insensitivities of the past ('leftist' errors), but now to 'take a clear stand against division and resolutely ... crack down on activities designed to split the motherland' (*Beijing Review* 16 May 1988: 15). This again signals the fundamental continuity in policies of the regimes before and after 1976 in their common determination to preserve the state's unity.

CHINA'S GLOBAL AND REGIONAL ROLE

I have shown how a geopolitical perspective helps the understanding of a variety of issues affecting China's internal and external security, including the determination to achieve agreed boundaries and the re-integration of territories. However, I have also emphasised that a geopolitical perspective on these issues is a partial one and that to understand them fully requires an appreciation of a much wider range of internal and external factors. Here I sketch in aspects of China's role since 1949 as a global and regional power so as to put the more immediately geographic questions discussed above into a wider context.

The introduction to this chapter emphasised that the traditional perception of China as the central kingdom had been shattered by the late nineteenth century. When Mao announced in 1949 that China had 'stood up', he also saw China in a threatening international environment, located between two superpowers. The United States was not only firmly established as a pre-eminent force in Western Europe, it had effectively become a Pacific power. American forces were secured in a variety of bases in the western Pacific. Japan, Taiwan, the Philippines and South Korea were forward projections of US political control and economic interests. The Soviet Union had gained domination of the buffer states of Eastern Europe, and with Japan defeated it secured its role as an Asian power.

The new Chinese government found its overtures for peaceful co-existence rejected by the United States. Caught between the two super-powers, it 'leaned to one side' and in 1950 signed a twenty-year military and diplomatic treaty with the Soviet Union. Though China professed an independent role and emphasised its links with Afro–Asian states, in the 1950s it remained subordinate to Soviet power. In the western Pacific and Southeast Asia, China was confronted by American power as successive US administrations pictured China as expansionist and bent on achieving world revolution. The USA sought 'containment' of China by establishing military bases and alliances to its east and south.

In the 1960s the Soviet Union and the United States moved towards implicitly recognising each other's sphere of influence – for example a basic territorial accommodation was reached over the division of Europe. As China failed to gain Soviet support for its own geostrategic concerns, particularly over Taiwan, so Sino–Soviet relations deteriorated. In 1969 they clashed militarily over their boundaries and the Soviet Union greatly increased its troops in Mongolia and the Soviet Far East. China had become embroiled in a geopolitical nightmare, sandwiched between two hostile superpowers who were simultaneously threatening its territory. That China became a nuclear power in 1964 gave it a certain security against both Soviet and American power.

In analysing contemporary Chinese foreign policy the critical turning point is not the death of Mao but President Nixon's visit to the People's Republic in February 1972. Clearly Mao Zedong was central to this invitation, which ended China's perilous encirclement. For the United States it was an implicit acceptance that Vietnam was not a strategic priority and a recognition of limits to its forward role in the western Pacific. As the United States and the People's Republic came into closer accommodation, China gained more access to Western technology and capital (see Chapter seven). In effect, China was now leaning to the side of the United States, as relations with the Soviet Union deteriorated and it perceived Soviet power in Asia threatening to contain China. In 1979 China clashed with Vietnam partly because of the latter's role as a projection of Soviet power in Asia. In the same year, the USSR increased its combat readiness in the Soviet Far East, and by invading Afghanistan extended its chain around China while threatening its ally, Pakistan.

Since 1979 China has attained a more independent role in the triangle of relations with the USA and USSR. This reflects a determination not to be subservient to any other power and displeasure at President Carter's 1979 agreement to supply Taiwan with 'defensive' arms and the Reagan administration's continued support for this. It is this context that helps explain the cautious moves towards better relations with the Soviet Union over the border and other issues.

Seeing China's foreign policy as part of the interactions of such a power triangle helps our understanding of major issues. However that analysis needs to be complemented by recognition of Japan's increased economic and diplomatic role in Asia. Furthermore, one must recognise the limitations of China's global role and its ambitions as a regional power. China professes itself not to be a superpower, and clearly it does not (yet) have the economic and military capability to act globally.

Despite its size and vast population it remains a desperately poor Third World state. Its armed forces may be large but they can as yet only effectively operate at a regional level. The conflict with Vietnam showed their limitations in equipment, leadership and combat experience. Since then there has been a substantial reduction in the numbers of troops, partly so as to spend more on new equipment to modernise its armed forces: defence is one of the Four Modernisations. More high-level technology is being directed towards the army, and the navy is being equipped beyond its previously purely coastal defensive role.

In the 1980s China's nuclear weaponry moved beyond a limited regional capability when it successfully tested an intercontinental missile and a submarine-launched ballistic missile. Thus its security concerns, which are still limited to regional interests in Asia, are capable of being supported on a much wider scale. The effect of this may include an encouragement to the Soviet Union and India to reach agreement on delimiting their boundaries with China.

CONCLUSION

In analysing the security concerns of any state, considerable care has to be exercised in using a geopolitical perspective. It is easy to slip into geographically deterministic explanations. Instead, the complexity of factors shaping the security of a state, and how it acts to maintain that security, need to be stressed.

The value of a geopolitical perspective is that it does emphasise that questions of location, distance, topography and ethnic geography shape the security of states. In using this perspective to analyse contemporary China one can see a certain geopolitical consistency in the security issues facing China's leaders. Chinese governments since 1949 have faced inherited problems of the relations between 'China proper' and 'outer China', poorly defined boundaries and long-term changes in the relative power of states in East Asia. Moreover, we can see certain basic, national policies consistent with these issues – a determination to maintain a unitary state, to establish new, equal treaties, to delimit disputed boundaries and to maintain superiority in the face of other regional powers. In these areas there are basic continuities between Maoist and post-1976 Chinese governments. A geopolitical perspective helps to explain that continuity.

REFERENCES

Chang, D.M. (1986) *The Sino–Vietnamese Territorial Dispute*, New York: Praeger.

Rhee, S.M. and Macauley, J. (1988) 'Ocean boundary issues in East Asia' in D.M. Johnson and P.M. Saunders (eds) *Ocean Boundary Making*, London: Croom Helm.

Schram, S. (1969) *The Political Thought of Mao Tse-Tung*, Harmondsworth: Penguin.

Schram, S. (1974) *Mao Tse-Tung Unrehearsed*, Harmondsworth: Penguin.

Segal, G. (1985) *Defending China*, Oxford: Oxford University Press.

FURTHER READING

Dreyer, J.T. (1976) *China's Forty Millions: Minority Relations and National Integration in the People's Republic of China*, Cambridge, Massachusetts: Harvard University Press. Though now dated, it remains the most useful study of China's minority policy.

Grunfeld, A.T. (1987) *The Making of Modern Tibet*, London: Zed Press. A useful study, but it omits discussion of the impact of the Cultural Revolution, the period which probably created much of the resentment still felt in the region against Chinese domination.

Harding, H. (ed.) (1984) *China's Foreign Relations in the 1980s*, New Haven: Yale University Press. A useful collection of review essays that seek to assess China's (future) role as a global and regional power.

Kapur, H. (ed.) (1987) *As China Sees the World*, London: Frances Pinter. Analyses of international affairs by Chinese academics.

Prescott, J.R.V., Collier, H.J. and Prescott, D.F. (1977) *Frontiers of Asia and South East Asia*, Melbourne: Melbourne University Press. Detailed maps and commentaries on Asia's boundaries.

Segal, G. (1985) *Defending China*, Oxford: Oxford University Press. An analytical study of China's military strategy in various conflicts from 1949–79.

UPDATE

Beijing Review is a useful source for official Chinese statements. Western analyses that focus on military issues include the annual publications *Military Balance and Strategic Survey* (Royal Institute of International Affairs, London) and the *SIPRI Yearbook* (Stockholm International Peace Research Institute).

Postscript: the impact of China's 1989 crisis on regional development and spatial change

Terry Cannon and Alan Jenkins

This book was almost entirely written before the events of May and June 1989, and the callous and calculated repression of those demonstrating in Beijing and other cities. After such events, the book might seem rather insignificant or irrelevant. Why bother with a survey of the geographical connotations of Deng Xiaoping's leadership, when political and economic affairs seem far more influential in determining China's future?

Our answer rests in this book: the political crisis of 1989, and the economic problems which fed it, are inseparable from geographical issues. Spatial differences both affected the way the Dengist reforms of the 1980s operated, and those reforms themselves had their own impact on the country's geography, on the regional contrasts, the urban–rural differences, and the localism emerging from greater economic autonomy for lower-level administrations. In other words, the 1989 crisis was partly a result of the distinct ways the reforms took hold in different sorts of geographical space.

When people demonstrated for greater political openness and against party corruption, their actions (and those of the leadership which repressed them) were in some ways related to the political and economic use of space and territory. Income levels, and the right or ability to own resources (and have personal control over economic opportunities), varied considerably between city and countryside. Urban–rural differences seem to have been significant in the tensions of the 1980s. There have also been conflicts over the advantages given to some regions rather than others, especially the priority given to the Coastal Region compared with the role assigned to the West. Among the minority nationalities (many of whom live in areas included in the Western Region), many people regard the economic policies of the 1980s as being calculated to produce the maximum extraction of resources from their areas. Severe penalties have been imposed on them if demands for greater autonomy have been made; following the anti-Chinese protests of December 1988 and the use of armed force, parts of Tibet were put under martial law in March 1989, two months before Beijing.

Even at the local level, policies like the Special Economic Zones (SEZ) have national political significance and economic spin-offs beyond the small size of territory involved. The whole issue of regional specialisation and comparative advantage (of which the SEZ and coastal priority were part) created new tensions in Chinese society, as did the rural policies which gave rise to family farms, and village and small-town entrepreneurs. Those rural policies have also produced a section of the population which is free-floating, many of whom have found their way to an insecure, marginal and sometimes illicit urban existence. Some people from this group had a direct role in the protests.

In all, the reform policies of the 1980s made new and different uses of territory, resources and the location-specific characteristics of different regions and places, and gave rise to new tensions, as well as new opportunities. The crisis of 1989, and the subsequent reversion to central-ised party control of the economy, is a crisis experienced in and expressed through China's geography as an inherent part of the economic and political conflicts: while hardly uppermost in people's minds, geography had its part.

In this Postscript, we trace the trends and policies which seem to be emerging under the new leadership which established control during the events of May and June 1989, and which has since consolidated its control of party and government. Zhao Ziyang (who was Party General Secretary at the time of the events) appears to be in custody, and his position has been taken by Jiang Zemin. Although Deng Xiaoping's position has not been challenged (he was in any case 'paramount leader' without any formal government position), it is clear that many of his reform policies are now in abeyance, subordinated by the now dominant 'conservatives' to greater party control over the economy. The new policies which have followed the June events indicate (although they are still not clear or comprehensive as we write) some significant shifts in geographical priorities.

RECENTRALISING THE ECONOMY AND SLOWING GROWTH

What is clear is that the 1989 protests, whatever their virtues and faults, have put into power a group of party 'conservatives' who believe that they have the answer to both the 'reactionary' protest movement (which they consider was pro-capitalist and influenced by Western decadence), and to the complaints of the protesters about corruption and the distortions of the party's behaviour brought about by commercialisation.

In other words, the 'conservatives' have pushed the debate about political reform off the stage, and replaced it with the issue of how to run the economy. The *economic* complaints of the protesters, seen by the 'conservatives' as being in some ways quite valid, are to be resolved by recentralising the economy to remove corruption and speculation.

Arguments about democracy are side-stepped and deemed redundant, because they pose a challenge to the role of the CPC in fulfilling its duty to clean up the revolution.

What follows is an assessment of the leadership's policy responses which relate to geographical issues. It is necessarily partial and based mainly on news media.[1] It should be kept in mind that there is also a degree of uncertainty within the leadership as to what should be done, and unwillingness by some of them to undo all of the reforms. The new propaganda stress is on maintaining the reforms (including the 'open door' to foreign investment) and presenting them as acceptable socialist policies; at times it is as if they are trying to persuade themselves as much as outsiders.

CENTRALISATION AND PLANNING VERSUS LOCAL AUTONOMY

Four months after the June massacres Premier Li Peng made an important speech in which he argued that the main causes of dissent were corruption, unfair income distribution and inflation.[2] The medicine which his leadership prescribed was a deepening of the credit restrictions introduced to dampen down the economy in Autumn 1988, and a reintroduction of state control of key industries, with a return to a central planning system similar to that in operation up to 1978.

There were already signs that the credit squeeze was bringing down inflation in 1989. It was implemented by the imposition of strict limits on borrowing and a ban on lending by the hitherto permissive Bank of Agriculture. The latter measure in particular has had a drastic impact on the many millions of rural enterprises (manufacturing and services) which have sprung up in the last decade. These township and village enterprises (TVEs) have been reduced in number by about 3 million (from 18 million) in the last year or so, making an estimated 8 to 9 million rural factory employees redundant. Although TVEs are still likely to perform well (with economic growth at 15 per cent in 1989), the rate is considerably down from the 33 per cent average growth of the previous years.[3]

This trade-off between jobs and inflation has also affected the state sector, with a decline in total employment of 200,000 in October 1989 alone. The more numerous urban 'collective' enterprises (which employ nearly 35 million people) shed more than half a million employees.[4] However, in the replanned economy the government is likely to give preference to these types of enterprise in the allocation of credit and raw materials. Some of the new leaders would say that the job losses are part of the justification for recentralising the economy. They argue that TVEs have undermined the state and urban sector by drawing raw materials and energy away from them (by bidding up prices and through corruption), thus boosting inflation as well.

The restoration of a centralised planning system (which will involve major industrial and agricultural inputs and products, including coal, iron and steel, non-ferrous metals, cotton and grain) is intended to remove the dual pricing system. This arose with the combination of fixed prices in the state sector and competition (and hence market pricing – often for identical goods) in the commercial sector. With it comes a recentralisation of the allocation of investment funds, restrictions on local authorities' and enterprises' use of foreign exchange, and reduced autonomy for both state-owned enterprises and provincial/lower-level governments.

Investments in weak economic sectors?

There are other crucial components of the new leadership's policies directed at the structural weaknesses in the economy which were deepened by the reforms. In particular, the new government says it will increase investment in the very inadequate transport and energy sectors, and in agriculture. Transport and energy, despite many statements during the 1980s that they were to be rapidly expanded, remained underfunded. This seems to have been a result of the increased autonomy of enterprises and local government bodies, none of which had a commercial incentive to invest in these areas while there were more profitable things on which to focus.

Agriculture has had mixed fortunes under the responsibility system, although rural incomes generally have risen. Production expanded rapidly in the first half of the 1980s, but grain production slumped to 379 million tons in 1985 and has since fluctuated wildly, creating enormous difficulties. This has partly been a result of the market system and farmers' profit-seeking. In 1989 the level was back up to the 1984 record level of 407 million tons. But more damaging has been the neglect of rural infra-structure, as a result of the reduced role of the collective system. (Already in 1987 a system of compulsory labour contributions was supposedly introduced to deal with the upkeep of irrigation and other infrastructure.)[5]

In addition to this, the state now intends to control much more the type of crops grown, and to restrict the transfer of agricultural land to urban and other non-farming purposes (to counter the loss of fertile land, especially in the eastern areas around cities and towns which have expanded rapidly during the reforms). Another big push to higher rates of mechanisation is planned, something which has been held back by the small plots inherent in the rural responsibility system (see Chapter six). The agriculture minister has said that more central funding will be made available to promote the changes in direction the government wants. Much of the immediate emphasis is on increasing output (especially of foodgrains and cotton) on large areas of land in the environmentally-favoured areas of the east and south.

Certainly the new administration is sensitive to the need to increase agricultural production, particularly grain. Official statements stress the need 'to genuinely implement the policy of taking agriculture as the foundation ... and that we must give investment, credit, energy and raw material priority to agriculture'.[6] But such directions are very similar to many others made in the 1980s, and so we must question whether such a policy will actually be implemented. As emphasised in Chapter one, previous calls for priority to agriculture have found competition in the power of heavy industry ministries and top party officials, who ensure priority of funds for industry. Given the increased role of the CPC in economic decision-making and the party's close ties to the industrial sector, agriculture may still not receive significant state investment.

Problems of local autonomy

It was the increase in the economic autonomy of the provinces which produced one of the more curious anomalies of the reform system. Instead of promoting increased market competition, autonomy made it possible for provinces to set up protectionist barriers to trade. They could then establish or maintain favourable conditions for their own industries, even holding on to raw materials for local use which previously went to state-sector factories elsewhere. Recentralisation is designed to prevent this, and is likely to incur the resistance of some provincial leaders who benefited from their local economic control. But the Beijing leadership is now stressing that the health of the national economy is more important than the self-interest or narrow focus of local authorities. As an article in *China Daily* puts it: 'Disputes over natural resources are a major source of discord between central and local governments. It should be made clear that natural resources belong to the State and that the State has priority rights over them. Local authorities should never stress their own interests at the expense of the State's.'[7]

Although the provincial leaders may share a regret at the passing of this decade of local power, their interests may be too diverse for them to put up a united opposition. An article in the *Far Eastern Economic Review* suggests that 'provinces with a high concentration of state industry, such as Liaoning and Heilongjiang, welcome the renewed emphasis on public ownership because they have got more raw materials and funds. On the other hand, recentralisation will hurt places such as Jiangsu and Zhejiang, which have more collective and private enterprises.'[8]

Another factor which also contributed to the job losses mentioned earlier is the decline in consumer spending, which has affected provinces which invested more in producing the type of goods which boomed in output in the 1980s. The slump in spending is partly a result of tighter credit and declining real incomes. But it is also an outcome of the 'inflation effect'. When inflation was high, people bought consumer goods rather than watch their savings lose in value as prices rose. As inflation has declined in 1989, there is less incentive to put earnings into buying goods.

Regionalism and priority areas

From this differential impact of the recentralisation policies, it is possible to begin to put together a revised impression of the likely regional pattern under the new leadership. This is considered in terms of how the fortunes of different provinces and sub-province areas are faring. As will be seen, much of the situation analysed in Chapters two, seven, eight and nine is likely to change. It is helpful to use the division of the three 'macro-regions' (see Figure 2.2) which were the basis of strategic thinking in the Seventh Five-Year Plan (1986–90). Unfortunately, there is much less current information about the impact of policies on the Central and Western Regions; their prospects are inferred as far as is possible after a more detailed discussion of the Coastal Region.

CONTINUED PRIORITY FOR THE COASTAL REGION?

In the 1980s the already generally wealthier and more productive provinces of the eastern and southern part of the country were given priority for further rapid development. It was thought that such a strategy would both promote the growth of the inland regions eventually, and maximise the gains from greater foreign trade and investment (easier to achieve on the coast).

Since Zhao Ziyang (who was a major proponent of this policy) was removed from office in June 1989, there have been many government statements that the coastal emphasis and the 'open door' policy are to remain intact. But it cannot be the same: foreign investment and some aspects of trade have been upset by the government's repression of the protestors. Nor does the new leadership seem to want it to be the same. Li Peng's October 1989 speech included mention of the Coastal Region remaining a priority, but it was not stressed. It also came with a qualification: China welcomed foreign investment, but 'we must meet the overall requirements of the development of the national economy and the rationality of our economic structure'.[9] Mention of the need for 'rationality' in economic affairs is usually a sign that what is currently happening is not approved of.

At present, policy on the 'open door' is something over which there is disagreement amongst the leadership, and therefore less clarity about what is meant to happen. Regional priority for the coast has been linked very closely to the use of foreign investment and to export processing. But the TVEs of Jiangsu, Zhejiang and Guangdong provinces (which have been at the forefront of both domestically-oriented industrial growth and the export drive) are now suffering from the credit squeeze and recentralisation. So the counter-inflation and planning policies are likely to increase the balance of payment's difficulties.

Some recentralisation measures (particularly those relating to foreign investment and trade) have hit even independent-minded Guangdong, especially through restrictions on imports and product quotas. These measures, designed to improve the trade deficit, stopped some joint ventures from operating, including the Peugeot factory in Guangzhou (Canton) which had to close for two months at the end of 1989.

On the other hand, 'coastal' developments which seem to be linked to other priorities may continue to thrive. First, there is no immediate clamp on the Special Economic Zones, and this seems to reveal their continued function as bridges for promoting reunification with Hong Kong, Macao and Taiwan. Second, the older state-led, industrial-base coastal cities like Shanghai and Tianjin (like the industrial Northeast) may benefit from the redirection of state investment to their industries and infrastructure under the new planning arrangements.

TVE areas: difficult times ahead?

Looking at Town and Village Enterprises, areas which have greater access to official funds or capital from outside China seem to be managing better than those more prone to the credit squeeze and recentralisation. Hence, the Yangzi delta region, incorporating north Zhejiang and south Jiangsu (the rapid development of which is described in Case Studies 6.1 and 8.1) has been hardest hit. Growth in industrial output in 1989 was less than 3 per cent, compared with more than 20 per cent per annum in the previous four years.[10] By contrast, the rate of decline has been much less in Shanghai's rural hinterland, where many TVEs are linked with state-run or other 'official' factories in the metropolis. In Guangdong, and to some extent Fujian, the existence of investments by Hong Kong and other 'overseas Chinese' has apparently cushioned the recession. There have been closures of TVEs, and as in the lower Yangzi area enterprises producing consumer goods for domestic consumption have been hard hit by the decline in demand. Yet 1989 growth was still around 15 per cent.[11]

Guangdong is probably also benefiting from its previous considerably greater autonomy, which is more difficult for the Beijing authorities to reduce for fear of further alienating Hong Kong. In addition, the province's governor is on record as being determined not to let party interference disturb the running of business.[12] Party dominance over economic issues (even at enterprise levels) is being reintroduced by the Beijing leadership to help in the planned economy, but is seen by many of the reformers (including Deng Xiaoping) as being a major factor in the inefficiency of enterprises in the Maoist era.

Special Economic Zones and Coastal Open Districts

The fortunes of the five SEZs have so far been a curious amalgam. Their location in relation to Hong Kong, Macao and Taiwan indicates that part of their function is to help in the process of reunification, as well as economic modernisation. Outsiders' confidence that China's leaders are willing to co-exist with capitalism in the SEZs, eventually leading to a two-system economy in a reunified country, has been shaken badly by the nature of the new post-massacre leadership.

To limit this damage, some leaders seem willing to continue to promote the SEZ system. In January 1989 Jiang Zemin was reported to have 'personally directed the planning and development of the Xiamen Special Economic Zone, [and] told the foreign investors that China's SEZs have excellent prospects and great potentials'.[13] Linking himself to the SEZ policy in this way is certainly designed to boost confidence among Taiwanese (across the straits from Xiamen). They initially stayed away after the June repression; but it is curious that investors from Taiwan seem to have continued dealing with the (recently enlarged) Xiamen zone, despite the anti-communist fears which they saw confirmed by the People's Liberation Army action. Trade across the straits continues to boom as well, bringing more connections between Fujian province and the Taiwanese.[14]

Reports in the official Chinese media have tended to reassert the continuation of the SEZ policy, but somehow they have to be imbued with a sense of socialist purpose. The most famous zone is Shenzhen (adjacent to Hong Kong) which is now described thus: 'By the sweat of their brow, builders of this brand new metropolis have begun to harvest the fruits of their work'.[15] On a visit to Zhuhai (next to Macao), Wang Zhen (a Vice-President of China) 'urged the people of Zhuhai to adhere to the style of hardworking and plain living'.[16]

These attitudes, combined with the recentralisation policies and restrictions on the SEZs' ease of trade and access to power and raw materials, have thrown into doubt the government's real intentions. The downturn in economic growth has in any case affected Zhuhai, with estimates of 30 per cent of its factories having closed in late 1989.[17]

But policy seems as yet uncertain: the SEZs may be an issue over which the factional differences in the new leadership are exposed, with the economic reformers (like Jiang Zemin) determined to keep them going much as before, and the 'conservatives' pushing for them to be more strictly controlled and less obviously capitalistic. Their function in generating modernisation and growth in the country's inland provinces has also been re-emphasised, implying that the zones are considered to be self-seeking or at best linking up with only nearby hinterland areas.[18]

In September 1989, the 'conservatives' signalled their views. The governor of Hainan (both a new province and the newest SEZ) was sacked by Beijing, accused of using his power for personal gain. Since his offences were both commonplace and relatively minor, some interpret this as both a warning about corruption and as the removal of one of Zhao Ziyang's supporters.[19] However, it did not lead to a widespread purge of independent-minded officials in coastal provinces (such as Ye Xuanping, governor of Guangdong), perhaps indicating both the nature of the balance of forces in Beijing and the power of Guangdong.

As with the SEZs, the government seems to want to give the impression that the coastal priority policy is still intact, though there are indications that it is opposed strongly by some leaders and provinces. However, there continue to be announcements of new projects which derive from the coastal preference. For instance, Shandong province announced four new export processing districts based in coastal cities. These are to use locally generated capital, and have independent powers to handle imports and exports.[20] Their function is much more in export promotion than acting as a modernising influence.

Shandong is one of five Coastal Open Districts recognised in early 1988 (but *de facto* in existence much earlier), and has seen considerable growth of TVEs in a manner similar to the lower Yangzi and Guangdong. Since June 1989 it seems to have been able to continue promoting its developments involving foreign trade and investment, although little seems to have been said officially about these five areas as a component of a continued pro-coastal policy.

Traditional port and industrial areas: preferential treatment?

The Coastal Region includes a number of old-established industrial centres such as Tianjin and Shanghai which seem likely to enjoy mixed fortunes under the economic slowdown and renewal of central planning. They will have to endure some of the economic constraints imposed in 1988 and reinforced by the post-massacre leadership. But at the same time, with their higher concentration of state-sector industry, they should receive more assured supplies of raw materials without having to compete with the TVEs, as in recent years.

For Shanghai, the blessing of state allocations is tempered by the fact that the greatest growth in the city in the 1980s was in its own non-state collctives and TVEs; they are also hit by the recession and recentralisation, so reducing the city's own revenues and making it highly dependent on state-sources of capital. In addition, the reforms' impact in the 1980s generated considerable capacity in other provinces for industries in which previously Shanghai predominated, such as silk textiles and engineering.

In other 'traditional' industrial areas, like the heavy industry bases in Liaoning, the new role for planning is likely to improve prospects by favouring these state enterprises with more certain supplies of coal, energy and raw materials. The political bargaining power of this type of coastal area is likely to increase under this process. It is clear that despite the use of the Coastal Region in the Seventh FYP as a single entity for macro-planning purposes, it is much more diverse than it was before, with different types of economic focus each with its own interests to pursue and with varying prospects under current policies.

The decisions of foreign governments and companies, and international agencies will also affect the viability of the 'open door' policy and coastal strategy. China has an estimated US$40 billion external debt, much of it due for repayment in the early 1990s.[21] The government may try to reschedule some of the debts, so as not to impose too severe restrictions on capital available for investment. But the burden of increasing the earnings from exports and invisibles (tourism has collapsed) to make repayments will still be enormous, and falls largely on the Coastal Region. This problem coincides with a situation in which Western governments and companies are (currently) much less willing than in the 1980s to invest and grant credit; they may now consider that political instability makes it too risky.

The Central and Western Regions

Provinces which have a higher proportion of primary-sector economic activities are probably less drastically affected by the new policies than those which built up new TVEs and reliance on the manufacture of consumer goods. This situation applies to a number of provinces in the Central (e.g. Shanxi) and Western (e.g. Qinghai) Regions.[22] Because much of their economic activity is in the state sector, the Central and Western Regions are likely to benefit from recentralisation. However, as far as foreign involvement goes, one source suggests that concern and investment in remote provinces has almost completely dried up (although foreign investment has generally been very low in areas other than the Coastal Region).[23]

Another factor likely to affect them, especially the west, is that the central authorities are less likely to expect the provinces to be financially self-sufficient. The economic reformers were reducing the levels of subsidy to provinces which had budget deficits, expecting them to develop more commercial (revenue-generating) operations of their own. Conversely, the provinces which produce budget surpluses (and these are mostly in the Coastal Region and some in the Central Region) are going to be forced to contribute more to central funds. One observer suggested late in 1989 that there may be 'a major redistribution of resources. Under Zhao, coastal and southern provinces fared better in terms of allocations than their inland counterparts. That trend may be reversed, with Peking making a major shift of wealth and resources, from the south and the east to the north and the west, as well as from the localities to the centre.[24]

Certainly the lower levels of resource endowment in the coastal provinces means that central direction over their use may favour the resource-rich areas in the centre and west. In addition, a researcher at the State Planning Committee has suggested that the growth of TVEs 'in Central and West China will, for the first time, overtake that of firms in the coastal areas . . . but he predicted there will still be a large gap between the better developed East the the backward West'.[25]

Regional policy changes

Even the use of the three macro-regions themselves may be ended as a tool of strategic planning. A report from a Hong Kong source has stated that during initial discussions of the Eighth FYP in October 1989, the separation of the Coastal, Central and Western Regions is to be discontinued. In their place it was apparently proposed to use divisions into groups of industries which cross over 'regional' boundaries.[26] This is largely in line with central planning methods as used before 1978.

This policy change would indicate a major shift away from the idea of regional comparative advantage as a basis for determining the different types of economic activity to be pursued in different places. Such a shift need not signify any reduction in the actual dominance of the coast relative to the inland regions. But the coast might well receive less central funding for its infrastructure, hitherto seen as necessary for its advantages to be fulfilled. The policy change may also reflect the fact that some inland provinces' leaders were angry at the lower priority they were getting.

Non-Han minority peoples and the west

The regional priorities issue has also been a sensitive one in relation to the national minorities, which inhabit especially the provinces and autonomous regions of the west.[27] Ethnic conflict tensions have been high in a number of areas, despite the more relaxed attitude in the 1980s to minority religious and cultural practices. While martial law was lifted in Beijing in January 1990, it is currently still in force in parts of Tibet. Xinjiang saw large demonstrations in May 1989, some in support of the Tiananmen protests, and others in support of Islamic issues, including the demand for a ban on a Chinese book considered insulting to Uighurs.[28]

Economic policies affecting the access of minorities to their customary resources (land for herding in many areas, plus agriculture based on different systems to the Han) are as yet less clear. It seems that the new government in Beijing will continue to perceive the minority people's territory as a resource base for the needs of the rest of the country. In a visit to Xinjiang in late 1989, Li Peng hoped that the region 'will be built into one of China's major bases of agricultural products, providing more grain, cotton and sugar for the country'.[29] The central government also sees the area as a source of much more oil and gas, and recently verified new reserves in the Tarim basin have confirmed its major significance for the national economy. 'China is expected to invest up to 1.5 billion yuan in 1989–90 in a large-scale oil exploration and development programme in the basin.'[30] This source of ethnic tension seems likely to remain and perhaps intensify.

The (Tibetan) mayor of Lhasa sees the prospects for Tibet's development based on greater linkages with the dominant Han economy, including the promotion of lead and borax mines in the city's vicinity, and joint ventures with factories in China proper.[31] In Yunnan, recent propaganda also stresses the need for stronger connections with economically-developed provinces in the east. Successful links for a silk mill are mentioned, which have brought technicians in from Jiangsu and Zhejiang to help improve production.[32]

The authorities also seem to be paying more attention to rural poverty in minority areas. A recent report stated that there were 331 'poverty-stricken counties' throughout the country, of which 141 were of minority nationalities. This situation has led the new government to promote special policies for the minority regions. These stress the reduction of grant aid, putting more emphasis on support for existing economic activities to improve their returns and promote self-sustaining growth. But this approach includes explorations for the development of mineral reserves and other resources as a high priority.

As *Beijing Review* recently put it: 'Many people, in eastern and central parts of China particularly, have made remarkable improvements in their lives, while those still badly off, are found mainly in the western areas of China where there are highly compact minority communities'.[33] This renewed apparent caring for the worse-off areas may be partly genuine, partly to reduce ethnic tensions and partly recognition of the resources to be gained.

URBANISATION AND RURAL–URBAN RELATIONS

During the decade of economic reforms, it has been estimated that as many as 80 million non-agricultural jobs were created. About half of these were in TVEs, a further quarter in the state sector, with 11 million more in urban collectives and about 6 million in private enterprises.[34] Through this enormous expansion, a significant amount of rural surplus labour was absorbed, much of it going to jobs in small- and medium-size towns as well as the cities. As suggested earlier, many of the new jobs were in coastal areas, especially where the TVEs were successful.

A *Far Eastern Economic Review* analyst suggests that during the cooling down of the economy there has been a deliberate favouring of the state sector in protecting jobs. To minimise unrest among state urban workers, the government has concentrated on cutting rural industry and private business. So, it is argued, 'the government will be forced to impose controls on jobless rural workers from entering cities'.[35] It is not clear that this is an essential outcome of the job losses which have occurred in 1989, unless these people are also seen as potential protesters. How much more easily can they be reabsorbed in agriculture than in urban areas? The mechanisation of agriculture mentioned earlier as another government priority had better be capable of using more labour rather than less! As an article in a Beijing English-language newspaper put it '1990 will be a tough year when the surplus rural labour force finds it difficult to set up new firms or find jobs in the existing companies and some returned workers from bankrupt enterprises discover that the limited farming plots of their families do not need them'.[36]

Certainly state-sector industrial workers have been engaged in many protests and strikes against worsening conditions in recent years, and these seem to have continued in spite of the repression begun in June 1989. As the *Far Eastern Economic Review* put it: 'Worker unrest represents a more dangerous threat than student protests. Therefore, to limit the impact of austerity measures, the government has had to continue paying laid-off state workers rather than forcing them to find new employment.'[37] Events in Eastern Europe must also have played their part in the new government trying to mollify workers.

For those forced to look for new work in the rural areas, the new austerity may well deepen social conflicts over income inequality which were already very apparent under the 'get rich quick' policies of the 1980s. They also return to a countryside facing an increase in central authority direction of what they have to do (including possible compulsory labour on irrigation schemes and other infrastructure). This is a situation which is likely to be unwelcome to most peasant families. Guarding against urban unrest may only delay the social explosion, and shift it to the situation in the countryside.

CONCLUSION

Perhaps we are foolhardy to write this Postscript: by the time it is read, events will have led to further policy shifts and more changes in China's geography. Even as we write, a meeting of the National People's Congress (the 'parliament') is hearing speeches which suggest that the government is having to reflate the economy to some extent, largely in response to growing unrest intensified by unemployment, lay-offs and the closure of TVEs. (The same meeting is discussing the 1990 budget, and it seems that the increased spending on agriculture is to be only around 7.5 per cent, compared with a 15 per cent rise in defence spending.)[38]

We decided to include this section for two reasons. First, it is clear that – even if only temporarily – many of the reform policies of Deng Xiaoping's leadership are now in abeyance. This means that the book is a geographical analysis of 1978–89. Readers deserve a clear indication that the main chapters of this book, often written as though the 1980s' policies were to continue, should now be read in the past tense at least in some respects. Second, the policy changes and related spatial responses after June 1989 underline one of the main themes of this book: the overwhelming significance of government policies in determining the geography of China.

Our analysis of the events surrounding June 1989 has also shown the value of considering them from a spatial perspective. China's changing spatial organisation has its own impact on the policies of the state. This is not to argue that the spatial issues considered here directly caused the crisis of May–June 1989. The causes were many, and require analysis from cultural, historical, political, economic and sociological perspectives; these are beyond the concerns of this book. What we have shown here is that spatial issues were part of that crisis.

Recent government statements argue that the economic 'readjustment' is succeeding in 'cooling down the overheated economy'. Particular emphasis has been put on reducing inflation and controlling the money supply. The same source goes on to acknowledge that 'new problems have arisen, including a sluggish market, enterprises operating under-capacity and increasing pressures of maintaining employment'.[39] We have emphasised how central control has cut back rural industry: given the under-employment in rural China (described in Chapter six), this is likely to depress rural living standards, particularly in the small towns and in the rural areas which had been transformed into industrial outposts of the larger cities. In addition, rural areas seem to be expected to reaccommodate many of the millions who have sought jobs in the cities. Furthermore, if peasants are not convinced that increased production will increase their income and that goods they could buy from earnings are in short supply, agricultural production will remain sluggish. The availability of food may then remain uncertain and prices high.

In short, the regime faces a series of economic and social problems, while there are still urban demands for political reforms. The government is an uneasy coalition of economic 'conservatives' (who want to ensure party control of the economy), economic reformers who seek to ensure a role for market forces and enterprise autonomy (provided it is within bounds set by the state) and sections of the military leadership who had been willing to suppress the urban protesters. It may not be able to agree on how to tackle these problems. Furthermore, it is a grouping which is unlikely to survive Deng's death, at which time we can expect a power struggle in which the dominant faction is still likely to be that which is supported by the old guard in the military. Whatever policies are then applied will lead to a new round of changes in the geography of China.

UPDATE

See the sources listed at the end of Chapter one.

FURTHER READING

A number of journals have now included special editions analysing the policy implications of 4 June 1990. These include *China Now* No. 131, Autumn 1989; *Current History* September 1989; *Problems of Communism* September–October 1989; and *World Policy Journal* 1(4), Fall 1989. Others will no doubt follow.

NOTES

1. It draws mainly on Chinese official English-language sources, the Hong Kong magazine *Far Eastern Economic Review* and British newspapers. In contrast with the rest of the book, where we have not encumbered the text with notes, we have used them in the Postscript both to emphasise the understandable lack, as yet, of academic sources, and the potentially obsolete analysis we are presenting in changing circumstances.
2. *Beijing Review* 16–22 October 1989, p. 15.
3. *Far Eastern Economic Review* 8 February 1990, p. 47.
4. *ibid.*
5. *China Daily* 4 December 1987.
6. *Beijing Review* 19–25 February 1990, p. 16.
7. '"Economic separatism" emerges', *China Daily* 26 December 1989.
8. *Far Eastern Economic Review* 30 November 1989.
9. *Beijing Review* 16–22 October 1989, p. 14.
10. *Far Eastern Economic Review* 8 February 1990, p. 46.
11. *ibid.*, p. 46.
12. *Financial Times* 12 December 1989, p. 37.
13. *Beijing Review* 8–14 January 1990, p. 6.
14. *Far Eastern Economic Review* 16 November 1989, pp. 68–9.
15. *Beijing Review* 25–31 December 1989, p. 10.
16. *ibid.*
17. *Far Eastern Economic Review* 8 February 1990, p. 39.
18. See, for instance, the report in *China Daily* 29 December 1989.
19. *Far Eastern Economic Review* 28 September 1989, p. 10.
20. *Beijing Review* 25–31 December 1989, p. 36.
21. *Financial Times* 12 December 1989, p. 12.
22. *Far Eastern Economic Review* 30 November 1989, p. 69.
23. *Far Eastern Economic Review* 14 September 1989, p. 62.
24. *Far Eastern Economic Review* 30 November 1989, p. 69.
25. *China Daily* 29 December 1989, p. 4.
26. *Wen Wei Bao* (a Hong Kong paper regarded as close to Beijing) 31 October

1989, reported in *BBC Summary of World Broadcasts* FE/0603 B 21. It is also reported in Economist Intelligence Unit *China, North Korea country profile 1989–90* London 1990.

27. Although Inner Mongolia has been included in the Central Region of the Seventh FYP, much of what applies to the west and other minority areas applies also to it.
28. *Far Eastern Economic Review* 3 August 1989, p. 37.
29. *Beijing Review* 11–17 December 1989, p. 6.
30. *Beijing Review* 20–26 November 1989.
31. *Beijing Review* 28 August–3 September 1989, p. 22.
32. *Beijing Review* 20–26 November 1989, p. 35–6.
33. *Beijing Review* 25–31 December 1989, p. 29.
34. *Far Eastern Economic Review* 8 February 1989, pp. 46–7.
35. *ibid.*
36. *China Daily* 29 December 1989, p. 4.
37. *Far Eastern Economic Review* 18 January 1990, p. 17.
38. *Guardian* 22 March 1990.
39. *Beijing Review* 12–18 March 1990, p. 4.

Index

Note: arrangement is word-by-word, i.e. Mao Zedong before Maoming